# MIDGES IN SCOTLAND

GEORGE HENDRY

Cartoons by ***BAX***

MERCAT PRESS
EDINBURGH

First published April 1989 by Aberdeen University Press
Revised edition August 1989
Reprinted 1990

Reprinted 1993, 1994 by Mercat Press
Second edition published 1996 by Mercat Press
James Thin, 53 South Bridge, Edinburgh EH1 1YS



ISBN: 1873644 612

Typeset in Ehrhardt 10 pt at Mercat Press
from author-generated discs

Printed in Great Britain by the Cromwell Press,
Melksham, Wiltshire

# *CONTENTS*

# *ACKNOWLEDGEMENTS*

The first edition of this book became a best seller in Scotland within weeks of publication, to my surprise. It has continued to be bought and, I believe, has acquired a certain following. In the past seven years' research on midges in Scotland has once again been funded and together with advances overseas has contributed much to our understanding of midges and midge control. The pity is that we still know more about what chemicals kill midges than we know of their biology and ecology. This new account brings the midge story up to date.

Two editions and ten years' worth of advice, comment, debate and criticism make me indebted to a large number of people who have helped in diverse ways. Among these my gratitude goes (alphabetically) to Tim Birkhead, Martin Birley, Alison Blackwell, John Boorman, Art Borkent, Andy Evans, Bruce Halliday, Len Hill, Darrel Ho-Yen, Douglas Kettle, the late John Lindley, Chris MacKenzie, Angus MacRae, Bill Mordue, David Murray, Gavin Parsons, Eric Reye, Alastair Roberts, Carl Schreck, Mike Siva-Joty, Angus Stewart and Ashley Woodhouse. I have not acknowledged their individual contribution within the text—though they will no doubt recognize their part in this book. I also thank the many readers who have been inspired to write to me over the years; I value their interest. My thanks also to Tom Johnstone of Mercat Press for his patience. To Jenny, Hamish, Kato, Fergus and Ishy my love.

Balnacra, Ross-shire

# *PREAMBLE*

For years there has been a wee conspiracy that the midges should never be discussed in front of the tourists. By mentioning not a word about midges in the holiday-guides to Scotland, let alone giving simple advice on living with midges, in some way it is hoped that the midge problem will go away. Unfortunately, the midges will not go away and, year after year, the visitors are driven to distraction and even to dislike the Scottish Highlands. Fortunately many people do learn to live with the midges and to enjoy Scotland in all its seasons.

This account sets out to describe in fairly brief terms how, when and why midges bite, to say something about their biology and to consider ways of avoiding the worst assaults of a midge pack on your scent. Over the years many attempts have been made to control midges, both in Scotland and overseas, occasionally with considerable success, more often with unexpected costs. Generally, the midge has the last word in these conflicts with Man. At the very least then, this book will give the visitor and Highlander the satisfaction of knowing their common enemy. By understanding the ways of the midge, a lot more people should be able to enjoy the full splendour of the Highland summer without quite so many bites!

BAX
WHO'S THAT WITH JOHN BROWN?

# *INTRODUCTION*

The proximity of the biting midge and the wearing of the kilt gave rise to the Highland fling! This is probably the only joke that can be made about an insect which has long been a curse of the Highland summers. Like their relatives, the mosquitoes, it is the female midges which bite to secure a blood-meal for their developing eggs. Despite plagues of biting midges in the Canadian outback, in Siberia, on the sun-drenched beaches of Australia and the Caribbean, eminent travellers and experienced zoologists alike agree that some of the fiercest midges in the world are to be found in the Highlands of Scotland. Wherever they abound, midges cause discomfort, misery and some pain to warm-blooded animals including humans. Social barriers they ignore. Attacks are made on the high and low with indifference. It was biting midges which half-devoured Queen Victoria at a Sutherland picnic according to her diary of 1872 (perhaps in retribution for her visit to the memorial of the notorious James Loch, half an hour earlier).

Turning their attention to lesser mortals, midges were held responsible for serious delays in the building of the Krasnoyarsk dam in Siberia in 1960. Few insects have achieved front-page coverage in *Pravda* under the headline 'This evil can be conquered'! Back in Scotland, fifty years ago both the Scottish Tourist Board and the Secretary of State for Scotland, Tom Johnston, sponsored scientific research into ways of combating midges. The outcome was to show that there are several profound and costly solutions to the problem and just a few rather simpler answers. These simple answers rely on an understanding of the unusual lifestyle of the biting midge.

Man has dwelt in the Highlands for the last 8,000 years and has long learnt to live with all sorts of insect life and for most of that time without the aid of sophisticated insecticides and repellents. It is still possible to enjoy the Highlands without the support of these toxins and many people do. There can be few more extraordinary sights in the Highlands than to see travellers on the platform of Achnasheen Station waiting for the arrival of the 7.16 train on a warm summer's evening. Word has it that the local midges only get up for feeding 15 minutes before the train arrives. In that short interval they get to work to savage the tired, hot, defenceless traveller. But what of the friendly guard? He knows his midges too well and does just what his forefathers have done for generations. Come 7.20 p.m., the train draws away for Kyle, and the Highlands each summer night are given over to the midges. These creatures form a dominant and successful part of Scottish wildlife and by controlling Man's activities in the Highlands they are the guardians of some of the most beautiful and unspoilt areas of Northern Europe. The Highlands would not be the same without the midges.

## *HISTORICAL PERSPECTIVE*

The very word 'midge' is one originating deep in European prehistory, from the old Norse *my,* the *mygge* of the Swedes, the *mugge* of the Dutch and *mycg* of the Saxons. But despite its antiquity, the midge gets few references in the early literature of Scotland. Not one word is given by Tacitus in his account of his father-in-law Agricola's excursions into the Grampians around 80 AD. One thing is certain, those bare sweating thighs of the Roman Legions would have been sore bitten as they struggled through the wetlands of Strathmore on their way north. Nor did the English host at Bannockburn pause in their flight to describe what it was like to have midges nibbling away under chain-mail or surcoat on that warm June day in 1314. Midges indeed get little mention in the literature of the earlier centuries—particularly in Gaelic culture. Maybe this silence tells us something of our modern-day perceptions of insect life.

It is not until the eighteenth century that writers start to comment on midges in the Highlands. One of the first was Edward Burt, civil engineer to General Wade in Lochaber. He wrote, in the 1730s, 'I have been vexed with a little plague ... swarms of little flies which the natives call Malhoulakins ... being of a blackish colour when a number of them settle on the skin, they make it look dirty; there they bore with their little augers ...' Burt himself was on horseback while his road gangs presumably had to suffer *na chuileagan* without relief. Objectively he noted: 'sometimes when I have been talking to anyone, I have endured their stings to watch his face and to see how long they would suffer him to be quiet; but in three to four seconds, he has slapped his hand upon his face and in great wrath cursed the little vermin'.

Bonnie Prince Charlie, hiding in the hills above Glen Moriston after Culloden, may have escaped the Redcoats but not the midges—or as one contemporary account had it 'the evening being calm and warm we greatly suffered by mitches ... to preserve him from such troublesome guests we wrapt him head and feet in his plaid, covered him with heather where he uttered several sighs and groans.' Later in South Uist, the now less-than-bonny *Teàrlach* was in a terrible condition where 'the mitches devoured him and made him scratch those scars ... made him appear as if he was covered with ulsers'. The entomological experiences of the 15,000 Government troops also out in the same heather looking for the prince are, however, unrecorded.

By the 1850s the Highlands had been opened up to gentlemen

diarists, game fishers, artists and ultimately the royal family. C R Weld on a sketching tour was to write, 'talk of solitude on the moors!—why, every square yard contains a population of millions of these little harpies, that pump blood out of you with amazing savageness and insatiability'. Robert Louis Stevenson writing *Kidnapped* from sunny Bournemouth clearly had memories of these insects when he had David Balfour much troubled by clouds of midges minutes before the shooting of Colin Campbell of Glenure.

Back in 1713 an Essex minister, the Rev. William Derham, set down one of the first scientific descriptions of the midge, or nidiot as he called it. He knew it as a greedy blood-sucker 'endued with a spear'. Serious studies of the midge really began in the late nineteenth century in Britain, Belgium, Germany and Russia. By the 1930s much of the British midge fauna had been described in some detail in Edwards' blockbuster of a book *British Blood-sucking Flies* and later in greater detail in Campbell and Pelham-Clinton's classic account. (The second author was later honoured by the naming of a Scottish species first recognized in 1984 as *Culicoides clintoni).* In the years after the Second World War several more species were discovered lurking in the hills in Scotland including *C. scoticus, C. achrayi* (after Loch Achray in the Trossachs) and *C. duddingstoni* (after Duddingston Loch in Edinburgh). Today some thirty-six species of biting midge are known from Scotland (see the Appendix). However, as early as 1946 it was recognized that most of the attacks on humans were being carried out by just four or five species, of which one in particular was responsible for most of the trouble. This species enjoys the full scientific name of *Culicoides impunctatus* Goetghebuer, or more simply the Highland Midge.

In 1952 the University of Edinburgh set up a Midge Control Unit specifically to study ways of combating the Highland Midge. Under the direction of Dr (later Professor) Douglas Kettle, a great deal was learnt about the life of the midge in Scotland and this work today forms the background for our understanding of the creature. Combined with the efforts of scientists in North America, Africa, Australia and Russia much is now known about midges, about their life-history, the way they reproduce, how they find their victims, in short what makes them such vicious little biters. The story that has emerged is a fascinating example of dedicated study frequently conducted under difficult and often painful conditions out on the hills and moors. From 1960 sustained midge research in Britain lapsed for 30 years though in recent years it has benefited significantly from a welcome (but, sadly, short-term) support to the universities of Aberdeen and Dundee and the Institute of Animal Health which has extended our knowledge of midge biology using modern techniques. Ironically, privately-funded support has contributed more to midge research in recent years than industry, including the repellent industry. Overseas, new technologies are being explored to combat tropical pest species. One simple message has emerged from this research. Almost all of the advances made in tackling the midge problem owe their successes to basic field and laboratory research into the life-history, the population dynamics, the genetics and the ecology of the insect. Without a sound understanding of how and why midges behave the way they do, then all the chemical sprayings, repellents and biological controls in the world become a waste of effort and money. The pity is that this simple message has not always been appreciated. Tourist boards, governments, local authorities, hoteliers and the tourists themselves want action *now*. The trigger finger on the chemical spray-gun gets very twitchy. It is only when expensive eradication campaigns fail, often spectacularly, that the call for more research is heeded.

Modern Man's almost instinctive urge to reach for the spray-gun has proved, time after time, to be a costly mistake. Midges are insects and like all other insects as well as higher animals such as fish, birds or mammals, they can be poisoned if enough toxins are thrown at them. Any chemical used against midges will, directly or indirectly, affect other harmless or even beneficial forms of life. It took twenty years to recognize this. Rachel Carson's book *Silent Spring*, which exposed the catastrophic effects on wildlife of the unregulated use of pesticides, had a profound effect on governments and scientists alike in the mid-1960s, and today we have learnt to temper or suppress our urge to spray. The reward for striving for alternatives to spraying, particularly in North America and in Africa, in the war against the mosquito and the tse-tse fly, has meant that today there are much more effective ways of tackling insect pests, ways which are less costly in money and particularly in terms of impact on the environment. To appreciate how these modern approaches work it is necessary to understand something of the world of the midge. The accounts which follow can only be a summary. For those whose appetite is unsatisfied, the short bibliography at the end will provide an open door to a fascinating world.

## *DISTRIBUTION*

Biting midges are found world-wide and have exploited their peculiar talents over much of the land surface of the globe. Most exploit the advantage of diminutive size. Many species prey on other insects. Others have evolved to feed undisturbed on a wide range of warm-blooded mammals or birds. Just a few have acquired a taste for human blood. Some 1,200 species of the most important genus *Culicoides* have been described so far, principally from sub-Arctic and temperate Europe, north and central America and Australia. Related genera are known from the tropical Caribbean seashores to the heights of Everest where the 1921 expedition complained of being persistently bitten at 14,000 feet. It is highly likely that many more undescribed species of midges exist in the tropics. One ardent researcher working in the central american state of Costa Rica recently found over 300 new species of biting midge in just one season of trapping. In Scotland, what we lack in variety we make up for in numbers. Most of the thirty-six known Scottish species prey on cattle, sheep, horses and deer. A few confine their blood-sucking habits to domestic fowl and birds, and four or occasionally five species attack humans. Most of the Scottish species have a characteristic, often local, distribution. Some restrict themselves to farm-yards and dung-heaps, others to sea or loch shores, a few to salt-marshes, mires or woodlands. Of the most persistent man-biters, just one species, *Culicoides impunctatus,* dominates in upland Britain, where it accounts for more than 90 per cent of the recorded attacks. There has been some indication that the *C. impunctatus* population in Scotland may be a race distinct from the *C. impunctatus* midges found over the border in England and in continental Europe. If this can be proven it would go some way to explain why the midges of Scotland have such a vicious reputation. The Scots who live in towns or in the Lowlands have learnt to live with another species, *C. obsoletus* or the Garden Midge. There is nothing obsolete about the way this midge bites; it does so less painfully than its Highland cousin but it is infuriatingly persistent. This species can be found throughout Europe, Asia and is widespread in Canada and the United States. A rather more painful biter, fairly common on the Scottish salt-marshes and elsewhere, is likely to be *C. halophilus* (known as the beast of Arrochar), while on the croft another species, *C. nubeculosus,* will attack humans, particularly in stables, byres, fanks and around closely packed cattle. There are several other midge species which will have a go at humans from time to time but they are rarely much trouble. It remains true, however, that no natural habitat or location in Scotland can be considered to be entirely midge free. Biting episodes in the hills above 1,500 feet may be reduced, but in sheltered corries midges can be active at considerable elevations. Nor is water a barrier. No less than three species of midge have been recorded from St Kilda, lying out in the Atlantic nearly 100 miles from the Scottish mainland. Only in densely developed cities and towns are midges generally considered to be a trivial nuisance, though even there the Garden Midge can get a hold in town parks and golf courses during wet summers. The banks of the River Kelvin, half a mile from the centre of Glasgow, are a well known midge-infested haunt in high summer. But to put this into perspective—midges were there long before Glasgow! Midges, indeed, have been found preserved in 75 million year-old amber. In Canada, so far 18 species have been identified from amber—all recognizable as ancestors of today's midges. Midges did not become extinct, unlike the dinosaurs on which some may have fed!

## *WHAT IS A MIDGE?*

As members of the Diptera (the two-winged flies), biting midges share several features in common with their relatives the mosquitoes. Biting midges, however, are particularly small with a wingspan of usually less than 2 mm. The Highland Midge is diminutive with a wingspan of only about 1.4 mm, a characteristic emphasized by its Gaelic name *Meanbh-chuileag* (tiny fly) (and hence the Highland greeting '*Tha a' mheanbh-chuileag dona a-nochd*'!). Any doubts about recognising a biting midge can be quickly settled by looking for the characteristic blotches or dark-flecked spots on the membranous wings. These spots can be readily seen using a low-powered hand-lens and are important, at least to taxonomists, in distinguishing one midge species from another. For example, the distinctive pattern of six to seven separate blotches on the wing of the Highland Midge is quite different from the arrangement of spots on the wing of the Garden Midge.

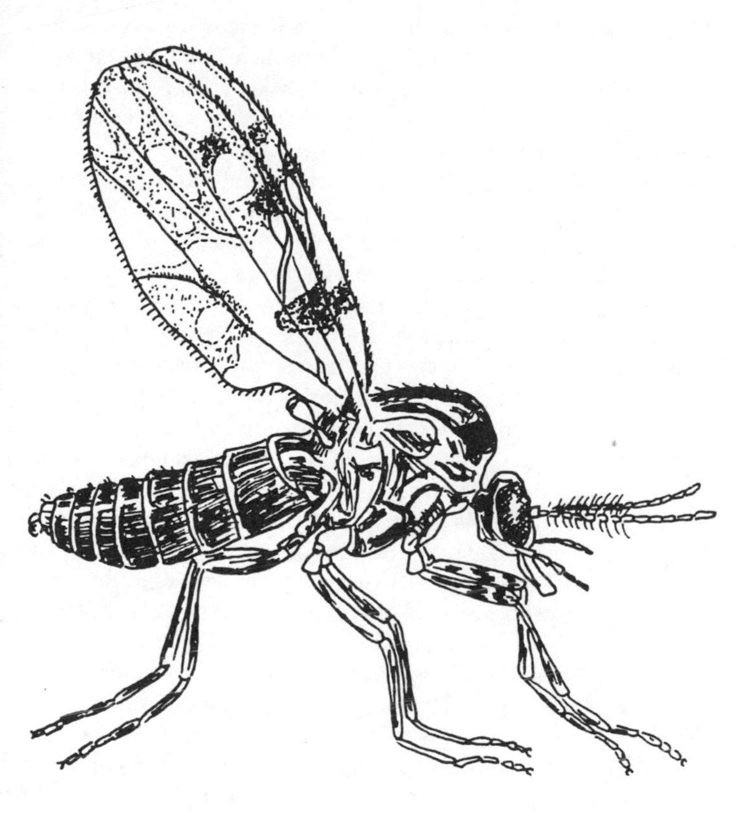

The Highland Midge in flight. Note the characteristic dark markings on the single pair of wings

If these dark patches cannot be seen on a living specimen then it is unlikely to be a biting midge. Once seen there is no mistaking it.

Another unusual characteristic of biting midges is the way they fold their wings scissor-like when at rest or when biting and feeding. The generally larger and more obvious dancing midges and related flies usually hold their wings roof-like when at rest. The most likely insect to be confused with the biting midge is the Chironomid or dancing midge. These insects, common in spring, summer and often well into late autumn, are significantly larger than biting midges, they don't bite and are most frequently observed as obvious, often dense, dancing swarms. There the insects, usually males, hover backwards and forwards for short distances but always facing the same direction, usually head into the wind. They will often occupy a favoured site for weeks at a time. If these dancing midges are displaced accidentally or deliberately they neither bite nor sting. Their sole interest in forming these dancing swarms is to attract females for reproduction. Instead they frequently get swatted in mistake for the real culprit, the smaller biting midge. Gatherings of biting midges do occur but they are much less populous and rarely obvious except to the trained eye.

Like other blood-sucking insects, the female biting midge has a well-developed, specialized mouth enabling her both to pierce the skin of her victim as well as to suck up the released blood. Piercing the skin is done by a pair of finely-toothed elongated mandibles and maxillae which work backwards and forwards with a scissor-like action, cutting ever deeper through the skin surface. When the cut or wound is deep enough a pool of blood is released from the fine capillary vessels underlying the skin. At this point the midge quickly rolls its mouth-parts into a fine tube or food-canal which is inserted into the wound to draw up the free-flowing blood. In mosquitoes and, in all probability, in midges too, saliva is pumped into the wound to prevent the blood from clotting and the flow from drying up. This saliva, with its digestive

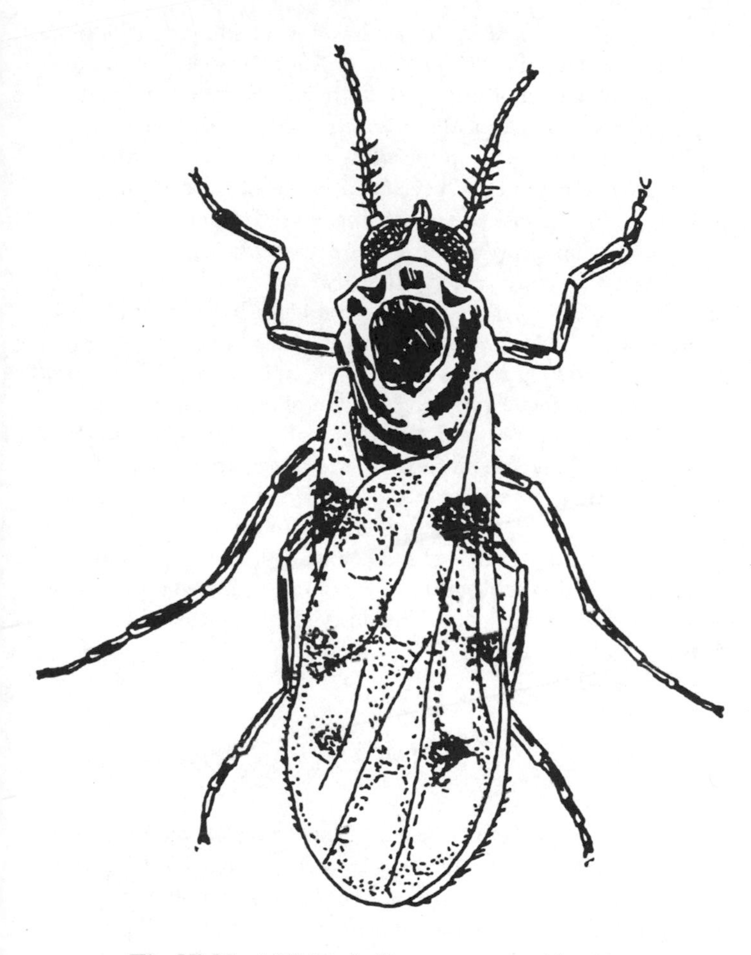

The Highland Midge feeding or at rest, with wings crossed scissor-like over the abdomen enzymes, now induces in the victims a mild allergic or immune response causing the victim, human or otherwise, to counter-respond with a rush of histamine to the site of the wound. One of the immediate effects of the histamine is to cause the wounded capillary vessel to open up for a few minutes to enable the body to mobilize and send in specialized white blood-cells which in turn set about the task of eliminating any infection and repairing the wound. Those few minutes between breaking into the blood-vessel and the body repairing the wound and staunching the blood-flow are a vital period to the midge. If left undisturbed, the midge will feed for up to 3 or 4 minutes before disengaging. Much of the itching and swelling of the wound area follows from the activities of the healthy body's natural repair machinery. The female midge has evolved its feeding habits to exploit the short time-lag before the blood-vessel closes down. It is a ticklish task for one so small. The male midge on the other hand has a simpler, less specialized, mouth; it doesn't bite animals and probably for that reason has been less studied scientifically. It is believed that the males derive all their food-requirements from plant nectar and from rotting plant remains. They are therefore of no direct trouble to humans.

## *BITES AND THE BITING HABIT*

One bite from a midge is insignificant, for most people passing unnoticed. However midges rarely work in ones. The evening stimulus for feeding affects large populations of pregnant and hungry females and it is this concerted and simultaneous assault, involving perhaps many dozens of midges, often concentrated over 20 or 30 minutes, which drives grown men to distraction and to seek shelter or relief from the maddening pursuits of an obsessive but unseen tormentor. Despite this, a number of dedicated scientists have sat it out, and methodically recorded midge bites at the rate of 2,000 to 3,000 bites an hour. By this means a fair amount is known about midge behaviour.

If it is any consolation, analysis of the engorged gut contents of the Highland Midge show that the primary sources of blood-meals, by far, are cattle, deer and sheep. Humans may be no more than a relatively hairless tit-bit.

What attracts a midge to its victim has been the study of recent research in Scotland. Cattle (and probably most mammals) release a complex alcohol, octenol, in their breath and this, combined with carbon dioxide, acetone, lactic acid and water vapour exhaled in breathing acts as a potent cocktail of attractants. In addition the pregnant female midge releases her own attractant (known from mosquito research as an invitation or recruiting pheromone) which signals the presence of a suitable victim to other midges. This seemingly altruistic invitation may explain why midges work in large numbers and perhaps underlies the old saying—kill one midge and a thousand arrive for the wake!

Having first detected its victim mainly by smell (and temperature differences) the midge homes in and attempts to land unnoticed on exposed skin. The close approach is probably conducted using the insect's eyesight rather than its sense of smell. If the landing is successful the midge will often wander over the surface of the skin presumably searching for a suitable soft area before beginning the task of cutting into the epidermis. Sometimes several false or abandoned attempts will be made before a satisfactory wound is secured. So far the victim will have felt little or no pain. If the midge continues to escape attention, some 3 to 4 minutes will be spent feeding on blood.

It is early in this period that the victim becomes first aware of a slight but persistent stinging or pricking sensation. If the midge can secure its meal without interruption it will finally disengage its mouth-parts and retire, fully gorged, for a rest period. If, as may often happen, feeding is interrupted, then the midge will show its remarkable powers of single-minded obsessive persistence when it will return again and again until it has secured its full quota of blood. Less than 0.1 µl (one ten-millionth of a litre) of blood is taken, an insignificantly small amount for the average human adult with about 5.6 litres to spare. To most people the overall result is no more than a mildly irritating spot, which subsides within a few minutes. But in the smaller number of susceptible victims the spots may rise to resemble nettle-rash accompanied by a puffiness of the affected area, particularly if bitten on the forehead, eyelid and lips. How each person responds depends, in part, on how often he or she has been exposed to midge bites before. To first-time visitors to the Highlands, the response to being bitten is minimal, this blissful state lasting perhaps for three to four days. These lucky souls have an immune system which has not been triggered to respond to midge bites, or more probably to certain proteins in the midge saliva. Because they have no specific antibodies with which to recognize midge saliva, the bite has no noticeable effect. But, unfortunately, this state of innocence has to end; with increasing exposure to attacks, the body's immune system reacts. In the blood-stream white cells form specific antibodies which for ever more will recognize the foreign proteins in midge saliva and will trigger an immediate response to future midge bites. So, on the next occasion when the midges attack, there is a rapid mobilization of the defence and repair mechanisms that exist in every healthy person, leading to the destruction of the midge saliva contents and repairs to the wound. The natural and healthy response of the immune system is to pump histamine into the site of the fresh wound to prepare the way for the protective work to be done by the white blood-cells. The result is a wound which is rapidly repaired and where the midge saliva is promptly detoxified. The price to be paid for all this healthy self-protection is that the victim is left with a mildly uncomfortable to slightly irritating spot, the irritation lasting for some few minutes. Most people, visitors and locals alike, respond in this way, given reasonably good health. A very few people over-respond by becoming quite ill and have to seek prompt medical attention. Their immune responses are rather too active. However, these unfortunate souls are usually well aware of their condition, often because they know that their bodies over-react to bites or stings from other kinds of insects. (There is some evidence that high concentrations of vitamin B1 taken over a period of more than 30 days before exposure to midges may reduce the severity of some people's reactions to bites. Others claim similar effects from a high dose of brewer's yeast). Among the local population, including long-term settlers, many individuals, but by no means all, develop a much more controlled, slow and less severe reaction to midge bites. These are the fortunate, seemingly enviable crofters who appear to escape the attention of the midges. They don't. It is only that with years of exposure to midge bites their immune system is rather slow to respond. Bee-keepers with years of experience similarly show a sluggish response when they are stung by bees. If it is any consolation there are a great many Highlanders who to their dying day suffer just as much as the tourist after the first week of a holiday in the area. The difference, however, is that these Highlanders have a knack of predicting when and where the midges are going to be out. Forewarned they know what to do to avoid the worst of the attacks.

## *WHEN AND WHY DO MIDGES BITE?*

Seasonally, in Scotland, several midge species are active from April until October. These early and late-season fliers are rarely of much trouble to humans. The Highland Midge, *C. impunctatus,* first appears on the wing towards the end of May or in early

June. Most of this early hatch are non-biting males who have no interest in molesting humans. The first serious biting episodes really begin in mid-June and persist to varying degrees of severity throughout July and into August. The biting season lasts for about 12 weeks. Towards the end of August midge numbers decline rapidly, just as many tourists turn south. By mid-September the midges are no longer an intolerable problem. The attacks are less prolonged and rarely more than a slight nuisance, although in some years, given a mild wet late summer there may be one or two last-ditch assaults until the end of the month. The Garden Midge, *C. obsoletus,* lingers on into October in towns, and in sheltered or coastal areas. Its bite is much less irritating and may even pass unnoticed. For most people, from September until the following June, the midges can be quietly forgotten, allowing the Highlands to be enjoyed to the full. Seasoned visitors to Scotland know that the Highlands are often at their best before and after the midge season; then the visitors can enjoy their hard-earned holiday without fear of a bite. For those who prefer to visit during the summer months there is still much that can be done to avoid at least the worst effects of the attacks.

Few midges in Britain are active biters throughout the daylight hours. The aptly-named *C. heliophilus* (sun-lover) is an exception and prefers to bite around high noon. It is, however, much less common than the Highland Midge. Like most of the Scottish species, the Highland Midge is particularly sensitive to light and, subject to weather conditions, bites most vigorously towards the long evening twilight. The first stimulus for biting in this species occurs towards dusk as the radiation from the sun declines to below 260 Watts/m$^2$. (In Scotland full sunlight on a cloudless noon in mid-summer is about 800 W/m$^2$). As the radiation decreases and the light fades, so midge activity begins. And as the sun radiation declines further to about 130 W/m$^2$, as happens at sunset on a cloudless midsummer evening, this is the moment for clouds of midges to rise from the ground cover such as heather and bracken in which they have been sheltering. What often puzzles people is why midges so often bite earlier in the day. The explanation is that high unbroken cloud, even at mid-day, can reduce the radiation from the sun to below the critical 260 $W/m^2$. Thick low grey-white clouds will take the radiation down to even lower levels as will dark rain-clouds. It is this cloudy weather which brings the midges out to bite in the morning or afternoon. In the forests, even on a cloudless day, the deep shade stimulates midges to bite for much of the daylight hours.

The apparent association of midges with warm humid weather is more to do with the cloud cover and lack of wind than humidity itself. Midges will bite avidly on dry evenings as the sun goes down. Relative humidity from a dry 65 per cent up to a moist 100 per cent has little effect on the intensity of biting though there is some indication that midges fly higher when the humidity is high and descend closer to the ground in dry weather. But whether they bite on the ankles or face, both can be mighty irritating. Drizzling rain does not deter activity either. Indeed midges and drizzle often go together on the West Coast, if only because where there is drizzle there is cloud and cloud means reduced light intensity.

The Highland Midge is hardy; it appears to be indifferent to all but unseasonably low temperatures, intense biting activities being maintained over the range 6° to 18° C. However, wind does severely limit midge assaults. At speeds of 2.5 m./second (about 5.5 mph), that is, up to a slight breeze, midge biting continues unabated. But above this wind-speed they are forced to seek shelter usually on ground vegetation or, where available, on dark-coloured tree-trunks. Young conifer plantations, bracken and thick unburned heather appear to offer suitable sanctuary from the wind. But once the wind passes and calm airs prevail, clouds of midges can be seen rising in unison, anxious to get on with the task of securing a meal.

Taking all environmental components into account, the single

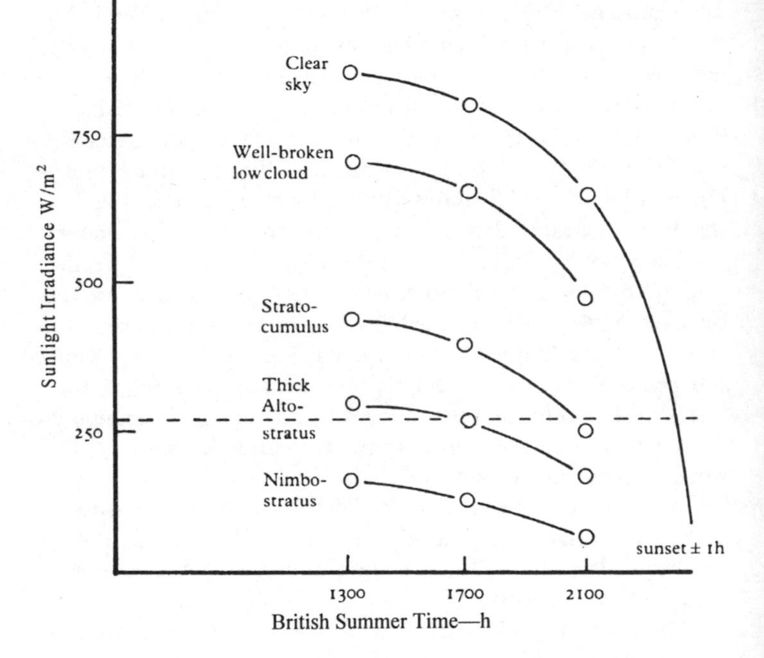


Sunlight irradiance through various types of cloud in the course of a mid-summer afternoon and evening in Scotland (latitude 60° N). The sunlight penetration is reduced by thick and low clouds and when it declines to below 260 W/m², this is the threshold for the Highland Midge to bite

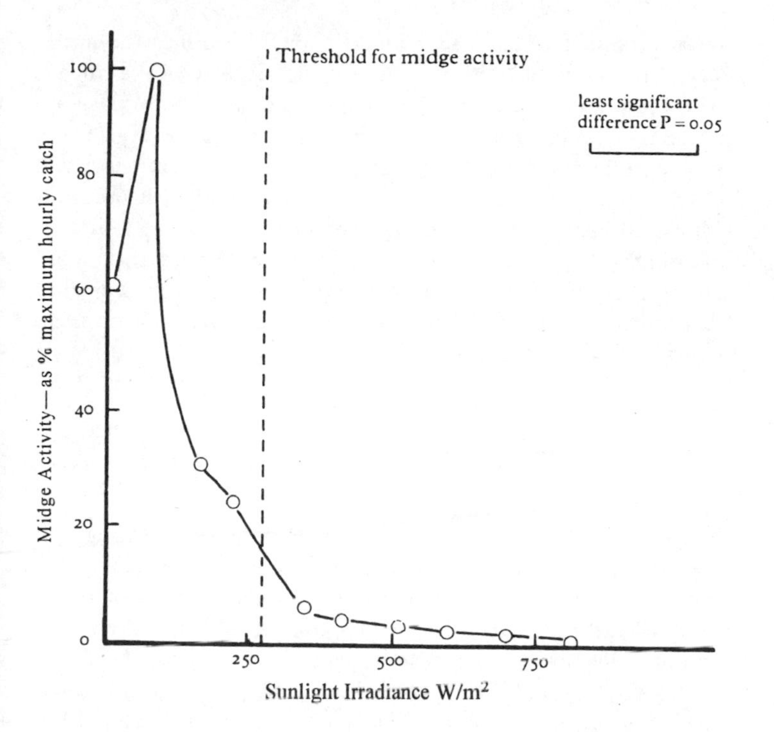


Flying activity of the Highland Midge in relation to sunlight irradiance. Above 260 W/m$^2$, sunlight suppresses midge activity. At about 100 to 130 W/m$^2$, midge numbers and biting reach their maximum most important factor which immediately determines the intensity of the biting activity is light. On cloudless days the midges will not stir until sunset. Dark rain-clouds and the midges are out in the middle of the day. In good summers, however, it is the prolonged evening and dawn twilight around mid-summer which offer an ideal environment for the Highland Midge. From mid-May until the end of July, twilight, at least the nautical definition of twilight, lasts throughout the night in mid-Scotland (latitude 56°N). It is no surprise, at least to experienced campers, to find that midges can and frequently do continue biting into the small hours and again before dawn.

## *OTHER BITING INSECTS*

Not every insect that bites of a summer evening is a midge. In Scotland there are other groups of flying blood-suckers, the Mosquito, Stable Fly, Black Fly and the notorious Cleg or Horse Fly. All are much larger than midges. In addition there is a motley collection of rather more personalized biters, the Fleas, Lice, Bed-bugs and Ticks, rarely seen perhaps, but all too obvious when active. There are other biting insects in Scotland but for most people they are not common. The account which follows is then a very brief summary of the most common Scottish biters likely to be met in midge country and whose bite could be mistaken for that of a midge.

It surprises many people to learn that mosquitoes are to be found in Scotland, though perhaps not in the numbers of their smaller cousins. Of the thirty or so species of mosquito in the British Isles, all are several times larger than a midge. These two-winged, delicate long-legged insects have a wingspan of about 4 to 5 mm (say ¼ inch), often with colourful iridescent or at least bicoloured bodies. Like the midge, only the female hunts for blood before developing her eggs. Unlike midges, however, most mosquitoes lay their eggs on or close to water, though this need for water can be met by puddles, rain-water butts or gutters on houses. In Scotland there are four types of mosquitoes, those that persistently inhabit buildings, those breeding in woodlands, those found on the sea-shore or on coastal marshes and, fourthly, those of the field and moorland. Predictably, those mosquitoes that share Man's dwellings or outbuildings will readily bite humans and domestic animals. Often known as house-gnats, many of these mosquitoes bite from April through to October.

Those species which hibernate over winter may give the odd mid-winter nip if disturbed. One member of this group of household mosquitoes was responsible for the malaria, or ague as it was known, that persisted in the fenlands of east England up to the 1850s and in north Kent to the 1920s. Migrant farm labourers from the Lothians took the ague back home from Essex every autumn. Malaria is extremely rare in Britain today. Despite their harmlessness, the female mosquitoes continue to inflict a painful bite, mostly at night and often when the victim is asleep. These house mosquitoes are readily recognized by their high-pitched piping or singing, particularly when flying close to the ear.

The forest-dwelling mosquitoes are perhaps less often met with; they tend to bite only near their breeding sites. The coastal species, sometimes found on shore-line seaweed, can be vicious and persistent biters. Perhaps the most common mosquito outwith buildings and away from the seashore is one with the name *Aedes punctor,* characteristic of moorlands, heath, birch and pine forests. This common Scottish species lays its eggs in dry holes or depressions that fill up with water in winter. The adult female is particularly persistent in her attacks when a suitable blood-meal moves into her breeding grounds. Fortunately mosquitoes in this country probably no longer transmit diseases; they are merely irritating. And on the west coast, mosquitoes are, at worst, a minor nuisance.

A quite different biter is the Black Fly. This short-legged plump black insect with a wing-span of about 5 mm is a scourge of Man in parts of northern Canada and Siberia. In several parts of Africa and Arabia these flies carry tropical diseases. In the British Isles one of the twenty native species is all too familiar for its unpleasant habit of feeding on the inside of horses' ears. At times they can cause considerable distress in stables. A related species of Black Fly is fairly common in the Scottish Lowlands and up to Strathspey where it will attack Man with a persistence which can be infuriating in the extreme. Once again, as with midges and mosquitoes, it is the female which does all the biting. The size of the insect, so much larger than a midge, makes its attention all too obvious.

For stealth, the Cleg or Horse Fly takes the prize. The stoutly-built, sombre, dusky-grey female, often working singly, alights so gently on its victim—horse, cow or human—to make a small painless incision in the flesh before slipping its mouth parts into the wound. It will even bite through clothing or hair. Left undisturbed, with its head down and tail cocked up, it will feed for one or two minutes. It has the audacity to make a second wound nearby if the first proves unsatisfactory. The pain comes when feeding is over, and the bite shows as a faint red ring which can persist and remain irritable for several days. There is often a distinct swelling around the bitten area. Clegs, unlike midges and many mosquitoes, feed mainly during warm bright days and, in Scotland at least, are less active in cool cloudy weather. In the Highlands they are only common in July and August. Clegs, mosquitoes and midges do share one feature in common; their attraction to dark-coloured clothing particularly when the wearer is perspiring profusely.

Scotland also has biters related to the African Tse-Tse fly. Known here as the Stable Fly, it is similar in appearance to the common House Fly. Both males and females will attack horses, cattle and, on occasions, Man. At times their numbers can reach intolerable proportions. The eggs are laid on decaying plant matter, particularly wet, soiled, animal fodder which has been allowed to accumulate in byres and stables. Persistent outbreaks of large populations of Stable Flies are often a sign of poor animal hygiene.

Bed-bugs, certain lice and fleas have long enjoyed human company. In these days of improved personal hygiene these parasites are less common than in previous centuries. However, in times of natural disasters, such as earthquakes or hurricanes, where access to washing facilities or a change of clothing becomes difficult, so these creatures multiply bringing with them outbreaks of typhus and, in times past, bubonic plague. Today, for most people, their presence is harmless and at worst an overnight irritant.

A more obvious irritant, particularly to bare-legged hill walkers, is the tick family. More closely related to spiders than insects, these round-bodied colourless to black or brown parasites rest on tall grass waiting for a bare leg to brush past. Having transferred itself to its new-found host, the tick sinks its head painlessly into the victim's flesh where it will hang for days, its soft body swelling to a considerable size before dropping off fully gorged. Young ticks may be no more than pinhead size and may be difficult to spot, particularly when they work their way into soft folds of the body. The deft way to remove ticks is with a pair of tweezers, held at the skin surface and twisted gently (twist not pull). If carelessly removed the pincers or proboscis may be left in and so introduce infection. A dose of whisky (applied to the tick) will usually encourage it to release its hold. As with their cousins on the continent and North America, Scottish ticks may carry Lyme disease and should not be ignored. Medical advice should be sought where a tick bite leaves a persistent red weal, a rash or a flu-like illness develops. Chronic Lyme disease can affect the joints, heart and central nervous system, so it can be useful to let your medical practitioner know if you have a history of tick bites. But again to put this firmly into perspective, crofters and keepers who work with sheep and deer are regularly exposed to ticks and have long learned to watch for wounds that do not clear up promptly. While Man has his share of annoyance from biting flies, his farm animals, deer, rabbits and birds appear to have more than their share. Apart from Horse and Stable Flies, many large animals suffer the almost constant attention of ticks, keds, warble flies, several species of black flies as well as mosquitoes and, invariably in Scotland, several kinds of midge. Although veterinary science has helped to reduce the scourge of warble and ticks in cattle, wild deer continue to suffer fierce attacks. Rabbits have their fleas, the vector for the myxomatosis virus.

There is some good news to be found in this unpleasant story, at least for humans. Away from byres, poultry sheds or stables, many of these biting insects are not common. Most are readily dissuaded from attacking humans by the application of products marketed as midge repellents. For most people, by far the most frequent biter in Scotland, particularly in the west and north, is the midge. With the possible exception of the Highland Midge, the Cleg and perhaps one or two species of mosquito, most flying blood-sucking insects show a marked preference for cattle, sheep, deer or birds. Only when disturbed from their natural or preferred host will they take to humans. So most summer visitors to the Highlands will rarely experience the attention of Stable Flies or Black Flies, although almost certainly a mosquito or two will make an evening attack. Visitors from the south, however, will experience far more bites from mosquitoes back at home. The average English cottage garden has quite a collection of mosquitoes and no holiday in Italy is complete without a fair number of mosquito bites. American visitors may be relieved to see the back of their mosquito population, though they may not be too enamoured with the Scottish midge. Clegs are occasionally a nuisance for perhaps two months of high summer in Scotland, but in south Germany and parts of France they are active biters from April to the end of October.

To put all this into context, many biting insects that are common in Europe, in England and North America have their counterpart in Scotland but in most cases in relatively low numbers. The distinctive ecology of the Highlands, the climate and particularly the high rainfall have tilted the balance in favour of just one group of biters, the midges. The rest of the biting insects seem to be, at times, almost a welcome conversational diversion.

## *BITING AND DISEASE*

Many important diseases are harboured and transmitted by biting insects, particularly in the tropics. These include malaria and yellow fever by several species of mosquitoes, sleeping sickness by tse-tse flies and river blindness by black flies. Concern about the role of midges as carriers of disease is natural and this possibility has been studied by scientists, medical entomologists and veterinary surgeons for many years.

The majority of biting midge species, including those in Scotland, feed on more than one host. The most commonly recorded victims are Man, his domesticated animals and many wild animals including deer, rabbits and wild birds. Subjectively, it seems that both the Highland and the Garden Midge prefer to attack Man but this may simply reflect the presence of human observers. Both of these species of midge are a plague on cattle, horses, sheep and deer. Despite their wide tastes, in terms of disease transmission to humans the Scottish midges are, as far as we know, harmless. Several species do, however cause extreme misery to horses, sheep and cattle. Deer herds both in Scotland and in Russia have been recorded as being driven out of low-lying pastures by increased midge activity and forced on to the impoverished slopes of the high hills. Midges may be one of the reasons why, until recent years, the Scottish Red Deer was an ill-fed and poorly developed specimen compared with its European cousins. Midges have long been blamed for distress in cows and in bad, midgy summers may be responsible for declines in milk-yields. Horses suffer greatly; a number of unpleasant equine disorders, such as sweet itch and fistulous withers are known to be caused by several species of midge. Some of these species show a marked preference for biting particular areas of the mane, belly or flank and to persist in feeding from the same wound. Much more serious, however, are the midge-borne diseases found overseas, particularly in the sub-tropics and tropics. There, several midge species transmit a number of serious viral, protozoan and microfilarial diseases to sheep, cattle and horses as well as to the indigenous wild animals. In recent years a number of outbreaks of these diseases have resulted in considerable economic losses to the farming communities in these countries and have provoked active veterinary research in the United States, parts of Africa, central America and in Australia. Outbreaks of one important disease in sheep and cattle, Blue Tongue Virus, have been notified in a number of Mediterranean countries in recent years. So far, there are no confirmed reports of this disease in Northern Europe. However it is worth noting that several common Scottish species will support, in the laboratory, the multiplication of Blue Tongue Virus (including *C. impunctatus*). African Horse Sickness Virus (carried by *C. obsoletus*) has reached Spain and Portugal through imports of Namibian Zebras, with lethal consequences to the Andalusian blood-stock. Relaxation of regulations in exchange of breeding stock within the EU, combined with more favourable climatic conditions, may see the spread of midge-borne diseases within Europe. The warning signs are there.

Humans appear to be more fortunate. Apart from a few debilitating disorders largely confined to the Caribbean and Central America there is no *hard* evidence that midges transmit diseases to human beings. Over the years midges have time and again been suspected of being a hazard to human health but have usually emerged with a clean ticket. In recent times there has been speculation that the AIDS viruses (HIV) might be transmitted by biting insects. Several authorities have examined this possibility but the consensus is that the AIDS viruses, almost certainly, are not transmitted by mosquitoes, close cousins of the biting midge. One simple observation is that AIDS itself or the antibodies to the virus are largely confined to sexually active adults and, with a few well-explained exceptions, are not found in children nor in the very elderly. Yet mosquitoes (and midges) bite the young, middle-aged and old without discrimination. Likewise mosquitoes and midges bite both sexes avidly. Indeed there is some evidence that certain midge species are attracted more to women than to men. The essence of these observations is that the pattern of the spread and development of the AIDS disease in human communities bears no resemblance to the biting behaviour and preferences of midges. This is particularly true in Europe where the disease is confined very largely to the male adult population. In addition to this circumstantial evidence, scientists have been unable to prove that the virus can even survive within an insect let alone be transmitted by one. Were it otherwise, the consequences would be too horrific to contemplate.

## *LIFE CYCLE—THE KEY TO SUCCESS*

One of the reasons for the success of midges in Scotland is their unusual life-pattern, one which has evolved to become well-adapted to the Scottish environment. Like most insects, the midge undergoes several distinct stages of development, from egg to larva to pupa, before emerging as the familiar flying adult. However, with midges, the egg and pupal stages are unusually short, generally lasting only for a few days.

Up to 10 months, however, is spent as a small maggot or caterpillar-like larva. And each species of midge has its own distinct or preferred habitat for its larva. While some species develop on the muddy margins of inland lochs, others exploit sea-coast marshes or farm-yards or drainage ditches. The Highland Midge larva grows from an egg on the blanket bogs, raised mires and poorly drained acidic grasslands that cover much of upland Britain, particularly in the wetter areas of the West. A suitable breeding site for this species is one which is relatively moist throughout the winter and summer but is reasonably free of standing water. Such sites are often marked by the presence of *Sphagnum* and *Polytrichum* mosses, particularly where there is some ground-water movement from the presence of nearby burns and ditches. To the botanist, the Jointed Rush *Juncus articulatus* and Sharp-flowered Rush *J. acutiflorus* are a good indicator of the larval breeding grounds of the Highland Midge. Just such sites are widespread throughout much of the West and Central Highlands and Hebrides. With some 4 million hectares at hand it is hardly surprising that enormous populations of midges can exist. Estimates of up to 24 million midge larvae per hectare (10 million per acre) have been made. It is likely that even these numbers are greatly exceeded in the coastal straths of the far West.

In whatever habitat the eggs are laid, a critical aspect of midge survival is the need for moisture. Both the larvae and pupae are susceptible to desiccation. A dry spring and early summer can spell disaster for the new generation of midges but for humans and livestock it can signal a relatively midge-free season. A wet spring usually means a midgy summer. West of meridian 4°30′, within 60 km of the Atlantic seaboard, the average annual rainfall exceeds 1,250 mm (50 inches) and in some places regularly exceeds 2,500 mm (100 inches). These high-rainfall areas in the west of Scotland are the heart of the midge country. In the drier east, midges become less of a problem. Much of the east

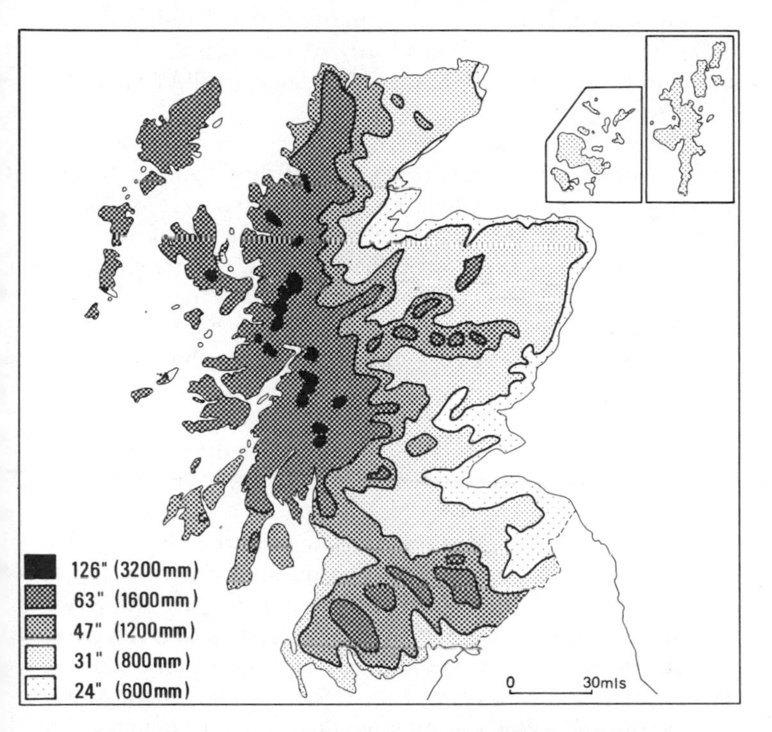


Average annual rainfall in Scotland. The Highland Midge is found predominantly in the areas of the west and north where rainfall exceeds 50 inches (1,250 mm) a year coast up to Aberdeenshire is comparatively free of midges though there are years when even the hardy Aberdonians suffer. Given the right habitat and sufficient rainfall, midges will breed by the millions.

The breeding cycle begins when the eggs are laid during the thirteen weeks of summer. In the case of the Highland Midge, the eggs are deposited three to five days after fertilization in batches of fifty or more, probably on the surface of saturated humus or on wet vegetation on the soil surface. After about ten to fourteen days the hatching larvae burrow into the top 5 cm of soil. There they develop, slowly, through four stages to mature, after Christmas, as a small, 5 mm-long, opaque-white caterpillar or maggot. While buried, the larvae acquire a voracious appetite and, during the ten months or so spent as a larva, the body-length increases some six-fold. Capable of limited motion, the omnivorous midge larvae rove through the moist soil feeding on a wide range of other larvae, nematodes, protozoa, green algae and, presumably, a wide selection of fungal spores and mycelia. Some species are not above a little cannibalism given the opportunity. When fully grown, from late April to July the larvae develop into the non-feeding pupal stage close to the surface of the soil. In some species, the pupae are capable of a flicking movement probably to counter the effects of desiccation. A prolonged drought at this critical stage may have a profound effect on the number of adults which finally emerge in the succeeding days. Good or bad midge-seasons, at least from a human viewpoint, will often depend on the rainfall in May and June as the pupae develop and mature. Prolonged dry weather has been shown to reduce populations of the Highland Midge by 90 per cent or more. Good, midge-free days during summer can often be traced back to a dry spell in the preceding month.

The first adults to fly, at the end of May, seem to be predominantly male. Several midge species form gatherings of all-male swarms usually 1-2 hours before sunset and persisting for perhaps

THIS IS THE LIFE, LADS -
WE GET FRUITFUL WHILE THE
LASSIES MULTIPLY

BAX.

5-10 minutes. They are rarely obvious, often short-lived and quite unlike the more familiar gatherings of the non-biting dancing midges. These male swarms, in the case of the Highland Midge, involve perhaps no more than a dozen or two of individuals which hover and fly backwards and forwards, spiralling up and down facing the wind, maintaining position over their chosen marker, often a puddle, an exposed rock or tree stump. Newly emerged females fly into the gathering where the distinctive frequency of the female wing beat probably acts as a come-on to the males. The females also release a scent which excites the males and encourages their mating attempts. The fine details of the mating process are now beginning to be recognized (just one benefit of recent research). Copulation lasts for 2-3 minutes initiated by a wrestling phase where the male gropes his way round to the underside of the female. *In copula*, as they say, the male's abdomen undergoes gradually strengthening contractions followed by a male-initiated separation involving much kicking. Once inseminated the female appears unreceptive to further attention—the semen may contain an anti-aphrodisiac, one way the male can ensure propagation of his genes rather than those of his co-swarming brothers. By limiting the frequency of copulation, in some insects and possibly midges, this helps to extend the female life span.

Either before or shortly after fertilization, the females search for a source of carbohydrates and for a blood-meal. The carbohydrates, for energy, would ideally come from nectar but in the Highlands, in June and July, nectar-bearing flowers may not be widespread, particularly over moorland. Other sources of energy are sought and may include sap-wounds from trees and rotting vegetation. But the critical step for the females is to secure a supply of fresh blood. For most midge species this is a key moment in reproduction; without it egg development is arrested and, given the short life-span of twenty, perhaps thirty, days for the adult, this is the point on which all future generations depend absolutely. So much is true for the greater majority of biting midge species in Scotland. However, the Highland Midge has one trick up its diminutive sleeve. Blood-meals are none too common in the Highlands. Ranging across the hills, glens and straths, humans and other large, warm-blooded mammals can be rare or even non-existent to an insect with a limited flight-range. The Highland Midge, unlike many other species, has partly solved this problem. It has evolved a way of maturing and laying at least some of its first batch of eggs without a blood-meal. Whether it sacrifices part of its flight-muscle, which carries a similar nutrient to blood, in order to mature the eggs is not known for certain. This process of maturing even a few eggs in the absence of a blood-meal (known as autogeny) is perhaps central to the success of the Highland Midge. At least it ensures that there will be some survivors for next year. To develop further egg-batches, however, a blood-meal is essential for this species. It is these females, having given birth once autogenously, which now form the bulk of the biting population. For good reason they are obsessive in their pursuit of blood.

Eventually, perhaps, a blood-meal is secured and at this point the females disappear (at least from the collecting records). Although it is not known for certain, it seems most likely that these well-fed females seek cover and rest in long grass, heather or on nearby tree trunks where they quickly lose the urge for further biting. They no longer react to the evening stimulus to attack. Over the course of the next few days, the eggs mature to be laid as before in a suitable damp breeding site. These eggs now represent the new generation which will emerge as flying adults the following year. Recent research suggests that eggs laid early in the summer may develop, given adequate rainfall, through the larval and pupal stages very rapidly, over a matter of weeks, to emerge as a later summer flush of new adults just in time for the start of the shooting season.

# *FLIGHT RANGE*

In open country, over grassland and moorland, midges will drift passively for considerable distances particularly in uninterrupted light breezes. Under the right conditions in summer, large populations can be carried 1,000 yards or more from the larval breeding grounds. Some hapless specimens have even been trapped many miles from land out in the North Sea. Others have been caught at considerable altitudes, from aircraft, in rising warm air-currents. The appearance of midges on some of the smaller West Coast islands unsuited to midge breeding, such as Iona and Staffa, may be due to drifting from the neighbouring mainland. This long-distance passive movement of midges in light winds is one important reason why chemical (or biological) larvicide spraying of the breeding grounds is ineffective in Scotland. It makes no sense to attempt to treat the open grounds surrounding, say, an isolated Highland hotel or caravan site when a nearby population of hungry females will be blown in on the next favourable wind current. Yet there are examples today of money being spent in futile attempts to provide an imagined *cordon sanitaire* round high-value sites.

In sheltered locations, particularly in established woodlands and in maturing forestry plantations, the adult midge population does not readily disperse far from the breeding grounds. Woodlands in the Highlands are often notoriously midgy, partly because of the reduced light from shading trees and partly because the still air in forests does not promote passive dispersal. Because of this the intensity of biting in a plantation is usually directly related to the nearness of the actual breeding grounds.

Away from sheltered sites, midge intensity diminishes with elevation. Just as red deer seek relief from midges by moving to higher altitudes, so hill-walkers know that it is possible to avoid midge molestation on hillsides by climbing ever higher. Indeed

FANCY HER GOING DRINKING AND DANCING IN HER CONDITION.

the relief to be gained from reaching an exposed windy site may be one good reason for driving the tired walker onwards to the summit. It is the prospect of a midgy night in the still air of the camp site down in the glen which spoils the idyll.

## *THE IMPACT OF MIDGES ON THE HIGHLAND ECONOMY*

The appearance of biting midges from late May to early September can be relied upon with confidence. Inevitably, this twelve to thirteen week period coincides with particular events in the Highland calendar. To the farmer and crofter in the West Highlands, this period is usually given over to the single hay crop and hay drying. This is also the period when much timber felling and extraction is carried out in the many forestry plantations. The summer months are often the most practical periods for road construction and repair. As autumn in the Highlands starts at the end of August and with snow falling as early as September or October, many trades and skills can only be effectively or reliably carried out during the midge season.

By far the greatest seasonal impact on the Highland economy is the welcome arrival of the summer visitors. They come from England, France, Germany, and in recent years from much further afield, with Scots returning from North America and Australasia, their arrival and departure being dictated mostly by national and school holidays. To adapt to this, many Highland hotels, guest houses, caravan- and camp-sites open for business only during the spring and summer. A number of restaurants, craft shops, museums and tourist information centres operate only in the summer weeks. A substantial element of the Highland economy thrives at precisely the period when midge biting

BAX

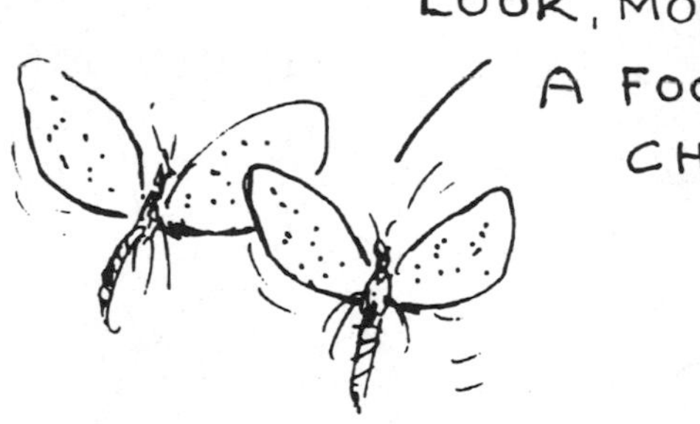
LOOK, MORAG –
A FOOD
CHAIN

NOW REMEMBER -
DON'T DRINK
THE DECOY.
BAX.

is at its worst. Given that the three months of summer can play such a dominant part in the local economy, it is surprising to find that little has been published on the impact of midges on the Highland economy. Market research and consumer surveys have been conducted from time to time but their findings have not been made public. As a result there are a number of anecdotal and unsubstantiated stories about the impact of midges on Highland life. For example a leading firm of Glasgow Chartered Surveyors advises clients to avoid putting their houses on the market during the midge season; would-be purchasers quickly lose interest if pestered by midges. A certain West Highland school had to discontinue running summer camp holidays because of the discomfort from midges. Deer and grouse shooters and game fishermen tend not to return next season if they have had to endure severe midgy days. Hotel keepers complain of cancellations following a sequence of prolonged midgy evenings though no doubt they benefit from the midge-bitten walker seeking relief in the hotel bar. But, away from the tourist trade, open reports are available on the impact of midges on other Highland industries. One report, for example, on the effect of biting midges on forestry showed that in the West Highlands, of the sixty-five working days of summer, up to 20 per cent of the time could be lost due to acute midge attacks on the forestry gangs. Of all occupations in the Highlands, forestry workers suffer worst from midge onslaughts. Hand-weeding and planting both involve working close to the ground vegetation and are notoriously prone to concentrated assaults. The work squads involved in timber harvesting have another kind of hell. There the men build up a fair sweat under their safety helmets and visors and create an odour that attracts midges in large numbers. As long as the chain saws are working the midges keep their distance, probably deterred by the exhaust fumes. But sooner or later the saws need refuelling or re-sharpening and this is the moment when the midges move in and drive hardened foresters to near despair. Indeed, there is an old saying among foresters that midges never attack a man standing with his hands in his pockets. But occupy his hands with a fencing maul, ditching spade or saw and the midges seem to know they can get a meal without being caught.

When the foresters return to their crofts in the evening, the midges are waiting for the intrepid soul who decides to cut or turn his hay. It is no laughing matter to see grown men raking the hay, with their heads and faces covered in their wives' nylon stockings. Ex-army respirators have at times been used for protection. Many other crofters are forced to stay inside on such evenings.

Forestry and, to some extent, crofting are part of the predictable future of the Highland economy. Both occupations have long adapted to the midge problem. What is less predictable is the return of the 13 to 14 million tourist trips made annually to Scotland, worth some £2 billion. Visitors to the Highlands spend almost £400 million each year in the region, the greater part in the months of summer. As these valued visitors are often not aware of the midge problem until it bites them, there is a strong case for ensuring that these welcome guests come fully prepared for the midges. That way visitors may be able to adapt quickly to the environment and enjoy their holidays to the full and to return in future years.

## *MIDGE CONTROL—IN FOLK MEDICINE*

Man has had something like 8,000 years in which to learn to live with midges in Scotland. From the earliest times he may have been responsible, unwittingly, for the spread of the Highland Midge. There is strong evidence that substantial areas of the Highland mainland, now covered in blanket bog, were once comparatively well-wooded and as a result well-drained. The repeated findings of preserved tree stumps at the base of the bogs together with layers of charcoal indicates that Scotland's earliest inhabitants began the process of forest destruction by clear-felling and by fire. The landscape, stripped of trees and opened to erosion in the high rainfall areas, developed into the mire formations of *Sphagnum* bogs so familiar today. It is just such areas that form the core of the breeding grounds of the Highland Midge.

Alongside their first pastoral activities, the early settlers must have quickly learned how to cope with the midges. One simple but effective discovery would have been the smoky bonfire. The smoke from burning wood and peat appears to interfere with the highly sensitive detectors on the midges' antennae and suppresses their ability to target potential victims. Smoke remains a useful deterrent to this day. The nearest equivalent to the portable smoky fire is the pipe full of tobacco, a device widely promoted for keeping midges at bay in several nineteenth century manuals on salmon fishing. The introduction of tobacco into Scotland in the early seventeenth century may have displeased King James VI, but by then he was down in London and safely away from the midge. In succeeding years the efficacy of tobacco found a new meaning in Scotland. In the late nineteenth century English workmen were brought to the island of Rhum, there to build the castle for the Lancashire industrialist Sir George Bullough. Despite their Accrington origins, the poor souls were required to wear Sir George's Rhum tartan kilt. The midges must have been delighted by this obliging exposure of prime English leg. Sir George nearly had a strike on his hands and only averted trouble by agreeing to pay his men a tobacco allowance of twopence a week. Although tobacco smoking is less favoured today, the resolution of ex-smokers has been cracked by midges in the past. Queen Victoria's water-colourist Carl Haag was given the task of painting views of Lochnagar, near Balmoral. Although this respected and stalwart Austrian had publicly given up smoking months earlier, it took just one really bad midgy afternoon to compel him to send his servant back for a box of cigars.

Herbal concoctions against midges have a long history. The common plant Mugwort (*Artemisia vulgaris*) owes its name to the old-English mycgwort, literally midge-plant. A related species, Wormwood (*Artemisia absinthium*) was recommended by Turner in his sixteenth century herbal as an insect repellent when used either as a lotion or when burnt on a fire. Both of these species are related to tropical plants which have long been exploited, commercially, for their potent insecticidal properties. The active ingredients are known as pyrethroids which in nature presumably act to deter insects from feeding on the plants. Some of these pyrethroids, particularly the modern synthetic ones, are toxic not just to insects but also to humans and should be handled with care.

Over the centuries a number of plant extracts or distillations to ward off insects have been described, some of which are potent midge repellents. By the 1920s several anti-midge nostrums had been published using oils expressed or distilled from a wide range of plants including lavender, lemon, lemon-grass, fleabane, wormwood, geranium, bitter orange, white cedar, cypress, eucalyptus, dill, fennel, juniper, tansy and thyme. Despite this long list, all of these plants have in common a number of chemically-related oils called terpenoids. Certain of these terpenoids such as limonene, pinene, thujone and citronella are proven and powerful insecticides. All have a strong and characteristic aroma and indeed several are widely used in perfumes, shampoos and soaps. Pinene, for example, is used in household floor-cleaners to impart a smell of country freshness. Citronella and geraniol are widely employed to give a lemon-smell to shampoos. Whether or not washing with such shampoos has any effect on midges has not been recorded. Thujone, found in Red and White Cedars (*Thuja* spp.), does seem to have a deterrent effect on midges; the foresters in the days before chemical repellents used to select a *Thuja* hedge or tree to have their lunch under as a midge-free

RIGHT, LET'S GO OVER THE PLAN OF ATTACK ONCE MORE
CASTLE ON ISLAND OF RHUM
BAX.

zone. But it was citronella which was most widely used as a pre-war midge repellent. The Forestry Commission issued gallon-jars of citronella to their workforce, right up to the 1950s. This had the reputation of turning the user an olive-yellow colour with continuous application. It has its adherents to this day and it can still be found for sale, occasionally, in Highland pharmacies.

Following the recent revival of interest in herbal alternatives to man-made chemical repellents, some of the old remedies are once again being offered for sale. One is marketed as a blend of natural repellents based on oils of citronella, pennyroyal, cajuput, lavender, bergamot and sassafras. Interestingly, the last component, sassafras, comes from a tree which used to be prescribed to ward off the effects of the ague (or malaria). Less exotic is bog myrtle oil, which has been recently marketed specifically for its anti-midge properties. Unlike sassafras, bog myrtle (*Myrica gale*) grows widely throughout the western Highlands and islands and has the potential for creating a little local insect repellent industry. The oil is a cocktail of volatile terpenoids (including pinene, limonene, eucalyptol with even a hint of citronella). Unfortunately, the commercial yield of oil is low and, as with many natural oils, if applied in too dilute a form it is ineffective. The catkin-like flowers have the highest concentration and there may be some benefit in applying crushed leaves and catkins, at least to old clothing.

In Australia where the coastal midges can be a serious nuisance, the Queensland sea-anglers have developed their own insect repellent. Having drunk the contents of a beer bottle or can, they half fill it with vegetable oil, add a quarter of a cupful of an antiseptic liquid such as Dettol, top up with water and shake vigorously. The foaming emulsion is then applied liberally to all exposed parts. Visiting midges may be deterred by the smell of carbolic acid or if not they will drown in the vegetable oil, it is claimed. The recipe has the advantage over most commercial repellents that it does not contaminate the bait nor dissolve nylon tackle. A less *macho* version uses equal parts of baby oil, Dettol and eucalyptus oil.

Each year a number of balms and costly potions are marketed as midge repellents but which seem to be better at attracting other humans than deterring the midge. However, the efficacy of one balm was discovered in the Florida Everglades. There, both midges and other biting flies are a serious menace to visitors. Some adventurous souls found that, by swathing themselves in a bath-lotion marketed as Avon Skin-So-Soft, they were able to keep the midge bites to a tolerable level. It may well be that small insects simply drown in the sticky lotion, though it is true that the ingredients include at least one compound which might well ward-off insects (as well as promoting soft skin).

In the scientific literature, there are regular reports of plants which are discovered to have potent anti-mosquito (and presumably anti-midge) properties. Most of these plants come from the tropics and warmer climates though many have relatives among our native flora or among our common garden plants. *Delphinium staphisagria* from the Mediterranean region is one. *Spilanthes mauritiana* from Kenya, a relative of *Helianthus* and *Dahlia* spp. contains a powerful mosquito larvicide. It would be an interesting and useful challenge to cultivate a midge-free garden in the Highlands by trying to grow some of these species here. For those gardeners willing to try, one group of plants worth considering for their anti-midge potential are those Compositae in the tribe Anthemidae (e.g. *Tanacetum, Artemisia, Anthemis, Santolina* and others ) which have long been known for their insectidal properties. With over 1,200 species there should be plenty to choose from. A midge-free garden in the Highlands would certainly make the headlines.

But perhaps a more simple way of coping with midges lies in another characteristic common to midges and indeed mosquitoes. These insects are attracted to dark colours. Scientists have long used black cloths laid over the ground specifically to catch midges alive, for further study. One explanation for this preference for dark colours is that many midge species take their rest on dark-coloured vegetation including tree-trunks. This ties in with the crofters' claim that the old breed of sandy-coloured Highland Cattle suffered less from the midges than did the Black Cattle. This is not an observation confined to Scotland. Experienced entomologists working in Africa often trap mosquitoes and midges using black cloths. In Russia, midges have been recorded searching for and landing on stationary dark-coloured cars, while avoiding light-coloured vehicles. Although proper trials seem not to have been conducted, this repeatedly observed preference for dark colours should not be ignored. It makes sense to wear light-coloured clothing whenever midges are about.

## *CONTROL—THROUGH CHEMICAL REPELLENTS*

It was with considerable foresight that the Secretary of State for Scotland, Tom Johnston, in 1944 asked his scientific advisors to find out if there was a suitable chemical repellent available which could be used against midges. Johnston was no doubt aware of the war-time campaigns fought against mosquitoes to control malaria and other diseases in the battle-grounds of south-east Asia. There, the allied forces were using various newly-released insect repellents, including one developed at the then Royal Technical College in Glasgow called dimethyl phthalate. DMP or Dimp, from army stocks held at Porton, became the subject of extensive field-trials in the Scottish Highlands. Quite what Johnston's motives were in promoting this research is not recorded. However, as an active and energetic founder and chairman of the then North of Scotland Hydro-Electric Board, he had first-hand experience of Scottish midges in the remote areas of the North-West. He may also have heard of reports that several army camps and naval bases in the West Highlands had become, at times, almost untenable in the midgy summers of 1943 and 1944. A mid-August walk today out to the old gun-emplacements near Aultbea will give some idea of the misery of sentry duty in those war years. The sites abound with midges which seem to choose the now abandoned buildings to shelter from the wind.

Johnston's scientific advisors, drawn from each of the Scottish universities, set to work in the summer of 1945. One team undertook a large-scale survey over 18 counties throughout Scotland, to determine the number of different species present and which ones were doing most of the biting. Nine thousand specimens later, the team identified one species, *C. impunctatus*, the Highland Midge, as responsible for 90 per cent or more of the trouble. This was enough information to set the second group into action. Based on headquarters at Raigmore Hospital in Inverness, field trials of the DMP repellent were set up initially at Achnashellach, Wester Ross, and at Auchterawe, near Fort Augustus, places deep in the heart of the midge country. With the co-operation of volunteers from the Forestry Commission, male and female, and with help from Land Army girls, roadmen, crofters, fishermen and schoolteachers from Skye to Sutherland, a methodical testing of DMP began. Within two seasons a satisfactory formulation had been developed and tested under West Coast weather conditions. Applied as a liquid or as a gel directly to the skin, or used to impregnate loose-woven veils and clothing, DMP rapidly gained approval. Its effectiveness, compared to the old pre-war repellents such as citronella, was quickly established and a Scottish manufacturer was found. From the late 1940s, DMP became available to the general public. Within a few years a second repellent appeared on the market, again from anti-mosquito campaigns, called di-ethyl toluamide or DEET. These two compounds DMP and DEET form the basis for most of the midge repellents sold in Britain today and indeed overseas. The repellents come in several forms. The most concentrated form is as a liquid, currently available at concentrations of up to 50 per cent or more. Higher concentrations become noticeably oily. Surveys among British and US forces found that DEET at 75 per cent concentration was sufficient to repel mosquitoes, higher concentrations offering no real advantage. Rather more expensive are the same ingredients packaged as aerosols. The concentration of repellent in the cans varies among the competing brands from around 12 per cent to 55 per cent, the difference being reflected in the price. As the propellant in these aerosols evaporates, what is left behind on the skin is 100 per cent repellent. However, in terms of value for money, most of the liquid preparations are a better buy than the aerosols. At lower concentrations, DEET is marketed in the form of gels and creams with the repellent concentration reduced to between 5 per cent and 35 per cent. For the seriously troubled, liquid preparations are probably the most cost-effective. Several of the repellent manufacturers claim that their liquid products offer up to eight hours protection. In fairness this assertion may be valid for the more sedentary user in cool temperatures but for those undertaking any exercise resulting in increased perspiration, the liquid can lose its effectiveness in one or two hours.

Although only the two compounds DMP and DEET are widely marketed as midge repellents, they come in many different brand names and prices. Among the more common ones are Autan, Boots Insect Repellent, Jungle Formula, Mijex and Secto-Combat (all containing DEET) or Shoo (DMP-based). Other brands imported from North America and occasionally sold by mail-order include another batch of DEET-based compounds such as Deepwoods Off!, Repel, Cutter, Muskol and Space Shield. Some products may additionally contain permethrin (more effective against mites than midges), citronella (perhaps for the pleasant smell) or a little glycerol to counteract any drying of the skin.

Apart from attractive packaging and presentation, the difference between the competing brands rests on price. Purchasers are urged to look at the price in relation to concentration. The price charged for very similar products can vary several-fold! Other so-called insect repellents are marketed from time to time, often with cosmetic or restorative attributes implied by the names 'milk', 'cream' or 'balm'. Usually based on ethyl hexanediol (or a similar sounding synonym), their effectiveness against midges is questionable.

Accepting that the most widely used midge repellents contain either DEET or DMP, how safe and how effective are they? On safety, both repellents are used in the plastics industry to soften hard plastics. The repellent manufacturers sensibly warn their customers to avoid bringing their products into contact with plastics, spectacle frames, polished or painted surfaces, rayon or acetate. Both DMP and DEET have been marketed as midge and mosquito repellents for nearly fifty years, throughout the tropics, in Australia and North America. Apart from the relatively small number of adults with unusually sensitive skins, there is no solid evidence to indicate that these compounds are not safe *when correctly and sensibly used.* Incorrectly used they are toxic and if swallowed can lead to acute poisoning. It has also long been known that many repellents, including DEET, are absorbed by the skin and reappear unchanged in the urine hours later. It makes sense then to treat all repellents with caution. For children, however, repeated and extensive application of even low concentrations of DEET can lead to a number of behavioural abnormalities. The British medical journal *The Lancet* published the following advice on the use of DEET (the journal has yet to consider the use of the other commonly available repellents based on DMP). The advice is as follows:

When used sensibly insect repellents are advantageous and safe, but the potential toxicity of DEET is high and the use of repellents containing more than 50% DEET should be avoided in infants and young children because of their thinner skin. Frequent total body application of DEET for days or weeks should be avoided.

These are sensible precautions which are followed by most users. One poor Finn, however, who applied 70% DEET daily to all his parts immediately before leaping into his open-air sauna developed acute manic psychoses in 14 days. Soldiers (presumably volunteers) who applied high concentrations of DEET frequently over many days developed skin rashes and blistering.

If these repellents were suspected as being unsafe in correct and sensible use, the various medicine and drug-regulating bodies throughout the world would long ago have called their safety into question. Instead these repellents are still, after fifty years, the standard authorized issue to the armed forces of the United States, Canada and India. Even so, fifty years of international acceptance of DMP and DEET-based repellents is no guarantee that these products are innocuous. Some manufacturers have recently reduced the concentration of DEET to 50 per cent following medical comment. A perhaps healthier alternative is to apply the repellents, not to the skin, but to clothing (provided not nylon-based). The clothing will eventually become stained but at least the repellent will not be absorbed by the skin.

Active research into devising new and safe repellents goes on in several overseas countries, usually targeted at particular types of biting flies. None have been approved for unregulated use by the public at large, at least not in those countries which control the sale and use of medicines. Certain of these compounds have been field-tested overseas, some against tropical species of midges. Whether they will ever be approved for more widespread use is uncertain, particularly with the hardening of attitudes on safety of insecticides in general. For the moment, most people are going to have to be content with the widely available DEET or DMP-based products.

There is, however, scope for improving the effectiveness of the currently-available repellents. In 1946, when DMP was undergoing field trials in the Highlands, volunteers were asked to wear loose-woven net cotton veils previously impregnated with the repellent. The responses were highly favourable but for various reasons, including perhaps the design or weight of the cloth, use of the veils was discontinued. More recently veils, or more correctly net jackets with hoods, have been manufactured and sold in the United States aimed at the hunter and game-fisher. These loosely woven, light-weight jackets come complete with press studs and draw strings. They need to be impregnated from time to time, perhaps once every few days, by simply storing the jacket in a specially designed bag containing the DEET-based repellent. They work by creating a repellent-laden aura round the wearer. Limited trials of the jacket have been conducted in Scotland, by coincidence near Achnashellach, the location of the original post-war midge repellent trials. One discerning Achnashellach resident, well known for his seven-acre garden, found that the midge jacket allowed him to work in the most midgy sites well into the evening. Without the jacket, some of his tasks had to be left over to the sunniest days when midge activity was low. A local crofter was able to turn his hay in the evenings, a task considered to be almost impossible after eight o'clock without the impregnated jacket. When freshly impregnated, the jacket is highly effective against the Highland Midge, it is comfortable to wear and does not appear to impede movement. Its drawback is that it is relatively expensive and could only really be recommended for those who are likely to be exposed to severe midge attacks throughout the season. A cheaper alternative is the midge hood; when impregnated this close-mesh veil completely covering the head will act as a physical and chemical barrier to midges. These hoods have found favour among the hill-walking fraternity. Problems over impaired vision have been overcome, though the issue of overheating on warm days has not.

INGENIOUS PROTECTION AGAINST MIDGES—A VALUABLE HINT TO SKETCHERS FROM NATURE.

(From *Punch*, 1853)

Wearing of the veil can become a choice between being hot and sticky or cool but bitten! A logical development has been to wear the head-veil but without any chemical repellent so relying only on the physical barrier. One company has recently extended this concept and now markets a complete head-to-toe suit of fine-weave nylon mesh, worn over normal clothes—the total barrier. An interesting revival of an old habit—Punch cartoons of the 1850s had visitors to the Highlands busily sketching Nature in head veils with holes cut out for the eyes and mouth!

Repellent-impregnated candles, coils, electric vaporisers and repellent strips are sometimes sold as midge repellents though it is doubtful if their effectiveness is anything more than extremely localized, though campers may find them helpful, particularly when combined with a fine mesh screen over the tent door. Ultra-violet lamps are also marketed which attract and electrocute flying insects. They are highly unspecific in their target and kill moths and many otherwise harmless as well as beneficial insects. They do not appear to make a significant impact on midge activity, however. Electric buzzers which claim to imitate the sound of male mosquitoes have been the subject of successful prosecutions under the Trades Description Act.

# *CONTROL—BY INSECTICIDES AND LARVICIDES*

After the launching of the first of the midge-repellents in Scotland in 1946, a scientific push was made towards tackling the major problem; how to eliminate midges and midge breeding from areas close to human settlements? At that time almost nothing was known about the ten-month long larval stages of the midge, so the first attempts to eradicate the insect were made against the flying adult. It was already established from work carried out in the United States and in the Pacific that mosquitoes could be controlled over large areas by spraying insecticides from aircraft. However, attempts to deal with midges in a similar way had proved much less successful, at least in the Far East. In addition, the economics of employing aircraft and specialized spraying equipment for no more than two months of the year in remote areas of the Highlands compelled the scientists to look at alternative ways of tackling the adult flying midge. Barrier spraying had been developed during the war to protect military camps from the nightly assaults of malarial mosquitoes. Strips of land 50 yards wide, frequently doused with DDT, had proved successful at keeping several mosquito species at bay. The decision was made to try out barrier sprays at selected sites in Scotland.

Several acres of moorland and woodland were treated with insecticides at concentrations known to kill most insects. The outcome was entirely unexpected. To all intents the adult midges appeared to be quite unaffected. As soon as the spraying operations were completed the midges emerged to resume their activities. It was not the DDT (or Dieldrin, Chlordane and Malathion also used for good measure) that was at fault. In laboratory tests midges, like most other insects, were readily knocked down by this highly toxic insecticide. From investigations in the field it appeared that midges were able to avoid all contact with the insecticide spray, probably by hiding instinctively on the underside of the vegetation. This ties in with the observation that midges at rest seek out dark colours and shaded sites. Similar results were subsequently obtained overseas during midge control programmes. From this it was clear that the diminutive midge had one up on its big cousin the mosquito. Barrier spraying was, then, spectacularly unsuccessful against the adult. However, DDT at relatively potent doses did destroy much of the midge larvae in the soil—along with much else of the soil invertebrate population. This near sterilization of the soil persisted for years, DDT being a notoriously persistent toxin. But, and this is the critical point, DDT and a range of now banned pesticides did not prevent midges coming in on the wind from nearby unsprayed sites.

If nothing could be done at the level of barrier spraying then the next target was the larval breeding sites. Apart from knowing little about the larva itself, next to nothing was known about where they developed. This gap in basic knowledge was a serious drawback. It was reasoned that if the breeding areas could be located on the ground then larvicides could be targeted to just those areas supporting breeding colonies. Identifying the precise location had another benefit; the toxic sprays could be restricted to areas away from livestock. Under Douglas Kettle (later to become a world authority on midges) a team from the University of Glasgow set out to locate the breeding grounds. Using specially designed traps which caught the newly emerged adult it proved possible to identify with considerable precision the breeding grounds of many of the Scottish species. Some species chose muddy farmyards, others preferred shaded woodland pools, freshwater ponds or stagnant mires. The Highland Midge, however, was found in none of these. Instead it was restricted to the damper areas of rough grazing and moorland, characteristically bearing *Sphagnum* and *Polytrichum* mosses and the jointed rush, *Juncus articulatus.* These plants now became the tell-tale for the otherwise hidden larval breeding sites. Closer examination revealed the small elusive maggots buried within the top inch of soil. These were the sites where the adult midges had laid their eggs in the previous summer and which now bore next year's generation in the form of larvae.

With the continued support of the Scottish Department of Health and with help from the Carnegie Trust and the Scottish Tourist Board, battle lines were set up in the form of a Midge Control Unit, under Dr Kettle, now based at the University of Edinburgh. Throughout the 1950s, a series of field trials were conducted on the hills outside Edinburgh and also at Talladale by Loch Maree in Ross-shire. Tests with several potent larvicides showed that the degree of effectiveness at killing the larvae depended on sufficient rain-fall to wash the chemicals into the soil. Without rain the insecticides were largely ineffective. The second observation was that once washed into the soil, the toxins were highly effective in reducing midge numbers. Rather more profound was the discovery that the larvicides persisted in the soil for several years. Unfortunately these persistent insecticides, widely used 40 years ago, had toxic effects on much more than just midges. The introduction of a chemical toxin into a food chain inevitably means that animals higher up the chain, including birds and mammals, will accumulate the insecticide from their diet. Because all animals, from insects to Man, obey the same biological principles, what is toxic to an insect is frequently potentially toxic to all higher forms of life if the dose is accumulated in proportion to the body-weight of the animal. The result of Man's profligate use of these insecticides and other toxic chemicals was graphically recounted by Rachel Carson in her best-selling book *Silent Spring*. As governments throughout the world began to recognize the devastating effects on wildlife by the late 1960s, attempts to control midge breeding by the use of broad-spectrum chemical sprays was largely ended, both in Scotland and overseas. In recent years rather more sophisticated chemical insecticides have become available which are either short-lived or claim to have a degree of specificity against one or a few groups of insects. These insecticides, it is alleged, have much less harmful effects on the environment. Several are used in seasonal combat against mosquitoes. In the United States, and elsewhere, local self-help organizations have been formed to control mosquitoes usually by methodically treating swamps and lakes by ground-based chemical fogging, or mist spraying, or by spraying larger areas from the air. No similar attempts have been undertaken in Scotland against the midge and the reason is simple enough. During the larvicide trials in Scotland in the 1950s, there was one quite unexpected finding which once again showed that the midge has one up on the mosquito. When an area had been thoroughly doused in the appropriate larvicide and the rain had washed the toxin into the soil the midge larvae were, of course, killed. However within days midge attacks were resumed. It appeared that adult midges from areas perhaps up to one mile away were moving in to bite in the sprayed zones. The only way to have eliminated these intruders would have been to extend the larvicide treatment several miles into the hills, an impossibly expensive operation. The monetary cost of such an operation confined to even the most severely affected areas would have been prohibitive. The ecological cost to Scotland's wildlife would have been unthinkable. For these reasons the use of larvicides against the Highland Midge, whether on a small community scale or over a larger land-mass, is not a practical solution to the midge problem.

## *CONTROL—THROUGH HABITAT MANIPULATION*

The problem of how to tackle the midge is not confined to Scotland. Many attempts have been made overseas, sometimes with unfortunate consequences, but occasionally with considerable success. For example in Australia the midge breeding grounds have been tackled fairly effectively by altering the physical environment of the site. In parts of Queensland some relief has been secured by clear-felling and back-filling the mangrove swamps, or altering the steepness of the shore-line and canal-side breeding sites. By damming or draining tidal inlets, it has proved possible to destroy midge larval habitats on a long-term basis. However, even these often costly landscaping efforts have occasionally back-fired. At one Caribbean paradise development, cursed by biting midges, the local hoteliers helped to finance the drainage of a nearby mangrove swamp, known to support a thriving colony of larvae. Having spent much money landscaping the reclaimed area, the developers were not pleased to find a new species of midge on the wing; one which was a more vicious biter and particularly active during the evening cocktail hour. The second species had previously been unable to establish itself in the area because of the too-wet conditions of the mangrove. But as soon as the site had been drained, the vicious biter found its own breeding paradise. Other developers have learnt from this experience and more successful landscaping has involved damming rather than draining the lagoons, so keeping a permanent stand of water between the midge and its chance to lay eggs. In Scotland, it is doubtful if any attempt to alter the landscape of localized areas would be successful in the long-term. There the Highland Midge appears to be more mobile than its Caribbean cousins and seems capable of drifting from distant breeding grounds right into the centre of towns and villages. Built-up areas in the Highlands such as Oban, Portree, Ullapool or Stornoway can be several hundred yards from the nearest extensive breeding grounds; yet, given the right wind conditions, midges can on occasions move into the centre of these small towns and make life uncomfortable. Only the most extensive programme of land drainage or water impoundment is likely to have any worthwhile effect.

Man has been actively manipulating the Highland ecology for upwards of 8,000 years. By 5,000 years ago Neolithic human activity had removed much of the original forest cover, such as it was. The later Holocene pollen records from 3,000 years ago show the extensive spread of moorland and acid grassland—linked perhaps to changing climates—habitats in which the Highland Midge would have thrived. The land-use history of the Highlands underwent significant changes with the expansion of cattle grazing (and droving) in the 18th century followed by the extensive introduction of improved breeds of sheep from the 1760s onwards, accompanied by abrupt displacement of long-settled human communities. The introduction of the percussion shot-gun and rifle after 1790 and the later popularity of deer stalking brought economically rewarding opportunities for extensive estates to be managed primarily for sport. The establishment in 1919 of the Forestry Commission and the widespread conversion of acid grassland and moorland to forest plantation particularly in recent years has created or enlarged new habitats. How much these changes have promoted the success of midges is a matter of continuing debate but it has been habitat manipulation on a grand scale. Intuitively, it is difficult to eliminate sheep as a factor—these four-footed, mobile, $CO_2$-releasing, compliant and relatively abundant sources of blood meals, conveniently shorn in June, diligently maintain and extend grazed and boggy grassland throughout the egg-laying months.

Given upwards of 4 million hectares at hand, it is extremely unlikely that the midge problem in the Highlands will be lessened by habitat manipulation on anything more than the most localized level. Queen Victoria could complain of being much molested by midges by Loch Maree in 1877—and that was seventy years before the forest plantations of Slattadale. Had she picnicked up nearby Glen Bianasdail her gracious words may not have been fit for even her diary!

# *CONTROL—FUTURE PROSPECTS*

Reviewing the past forty years of research into Scottish midges, the initial support for chemical spraying of adult midges or their larvae has now faded because the field trials proved to be ineffective or had unacceptable consequences for wild life. Despite these set-backs the research and field trials have provided much useful information about the way midges live. One benefit has been the development of repellents which, for all their defects, have offered considerable relief to the resident and visiting population. Knowing how to spot a probable midge breeding ground has enabled architects and builders to choose sites at least to windward of midgy grounds. Now that we begin to understand something about the life cycle and biting habits of midges, locals and visitors with a little experience can predict the likelihood of good or bad midgy days.

Left alone, the midges will continue to cause misery to visitors and locals alike. They will continue to disrupt human activity in forest plantations, in shepherding, stock-rearing and haymaking. They will displease hotel-guests, annoy game fishers and deer shooters and irritate hill-walkers, hitch-hikers and campers. On occasions, the midges will drive these good folk away with vows of never returning to the Highlands again. This is adverse publicity for the Highlands and it is, therefore, all the more strange that the agencies concerned with tourism should keep silent about the problem. Silence will not lessen the problem and lays the promoters of tourism open to charges of being less than honest. There is an alternative approach based partly on open discussion about midges and partly on tackling the problem by using the latest scientific developments in insect control. Although biting midges will never be eradicated from Scotland short of a major climatic upheaval, there are ways of reducing the midge problem to levels which are both bearable and which will allow the full benefits of the Highlands to be enjoyed throughout the summer.

Sustained research into midge control virtually ceased in Scotland in the mid-1960s. Overseas, research has continued, particularly in the United States, East Africa and in Australia, stimulated from time to time by outbreaks of midge-borne diseases in horses and sheep and by irate local authorities and tourist boards. In addition the governments of several developed and third-world countries support research into insect control for good military reasons. Midges along with mosquitoes and blackflies can severely impede the ability of ground troops to live off the countryside, whether in northern Canada or in southern India. All this has helped to fund research into new methods of control of the local species of midge. Scotland's midge problem, though acute by any standard, does not attract its share of research and development resources. Its small indigenous repellent industry together with that based in England and Germany has not produced any lasting, fundamentally new, synthetic products for over forty years. The prospects may be that nothing new will happen until someone tries out a product developed abroad against say, Venezuelan midges or the sand flies of Queensland. Such a trial would be a gamble for an entrepreneur; in the scientific literature there are examples of repellents which have proved to be highly effective against one midge species but much less effective against another. The last large-scale repellent field trials recorded in Scotland, conducted specifically against the Highland Midge, were carried out in 1947. The Forestry Commission has from time to time re-appraised new formulations from the repellent industry but always based on DMP or DEET.

Midge research is hampered not by lack of scientific talent but by lack of financial support. Since the early 1960s the Scottish midge has been seen as not a serious problem or, perhaps more accurately, accepted as a problem but one with no solution. (Had Tom Johnston thought in that way then even less progress would have been made immediately after the war). There are three areas in which co-ordinated research is required if the purely Scottish midge problem is going to be tackled. The first, and perhaps most wanting, is to establish the extent and cost of the midge problem to the social and economic fabric of the Highlands. Conducted by assessors independent of vested interest in seasonal employment or tourism and independent of the repellent industry, the survey would focus on a number of economic sectors. These would include hoteliers, tourists, static caravan operators, youth hostels (the impact on tourism), general practitioners and physicians specializing in insect-related disorders (the impact on health), forestry including the private sector, fish farming, the building trade, road workers, coast guards, electricity, transport and communication industries and crofting (the impact on Highland industries). Given this information, it might then be possible to direct efforts into particular sectors of the economy as well as identifying which industry or trade would support, financially, further research.

The second area of effort would be to assess the repellents and repellent technologies that have been developed overseas and to see if they are effective against the Highland Midge, and to compare these with some of the traditional remedies and natural repellents. What are called push-pull techniques are available—combinations of repellents (the push) with attractants (the pull) located at a distance need to be tested here. Modern repellent trials conducted under strictly defined, scientifically approved, criteria need to be carried out in the field under the varied climatic conditions of the Scottish Highlands. A number of tests have evolved over the years where groups of volunteers, male and female, young and old, are asked to apply a sequence of repellents usually to one arm only and then to subject themselves to foraging midges. Ideally the volunteers should be monitored while carrying out various tasks related to their normal occupations. To eliminate the likelihood that midges might be attracted to one individual more than another, each volunteer takes turn to test each repellent. Such trials, conducted by trained technicians under medical and scientific control, have frequently shown that one particular repellent works well under one set of weather conditions or is effective against one species of midge but is less effective under other conditions or against other species. So far as is known, rigorously controlled, extensive field trials against the Highland Midge have not been conducted in Scotland since

1947. The goal of such trials would be to identify which of the modern range of repellents, natural or synthetic, are superior to DMP or DEET against the Highland Midge on its home ground, bearing in mind safety and acceptability to the user. With sound medical and scientific backing, manufacturers or importers would be given the confidence to invest in new, safer and more effective ways of tackling the midge problem. While the immediate benefit would pass quickly to the consumer, the longer-term gain would fall to those concerned with the Highland economy, including rural industry and tourism. Having provided an opportunity to assess the social and economic costs of the biting midge and having secured at least a short-term solution through modern repellents, the third and most fundamental area of research would remain.

There is much that remains unknown about the ways of the Highland Midge. We still do not know how to culture the Highland Midge in the laboratory (that unfortunately takes time and money). If radically new approaches are to be sought for tackling the midge problem, then a new understanding of the midge is badly needed, particularly in the fields of insect physiology and behaviour. Some encouraging progress has been made in Scotland in the past few years, particularly in understanding reproduction and population biology, and in determining the role of smell in midge behaviour. This is a good start but this will be wasted if this groundwork fails to find further support. British science has all too often initiated important discoveries, opening the door to others to explore the finer details from which technological exploitation develops.

While we wait for this research we can begin to address the midge problem intelligently. For example, it is already possible to predict 'good' and 'bad' midgy days, at least on a local level. It would take no great stretch of the imagination to visualize a 'midge forecast' on local or regional radio or in the newspapers, in much the same way as pollen forecasts are made for the benefit of hay-fever sufferers down south. Forewarned, people might then be able to plan their day so as to avoid the worst of the assaults. One enterprising community radio station, Lochbroom FM, has, indeed, recently started midge forecasts. To put this forecasting onto a meteorologically reliable basis some fundamental field studies would have to be resumed from the point where they were left off in the late 1960s. We need to know reliably what are the consequences of dry or wet periods in spring and early summer on midge biting several weeks later. We still know next to nothing of the effect of climate on reproduction, the process of selecting breeding grounds and egg-laying. This is a critical point in the life cycle. After egg laying, does the midge immediately set out for a new blood meal or is there a delay dependent on weather conditions? What conditions promote the males to form swarms? What part does the weather play in the development of the mid-summer second generation of midges?

In the medium term, following the hesitant funding of recent years, exciting new developments are possible. Now that we have the technology to study the specific body odours that attract midges, can we humans interfere with this, perhaps with a counter-odour operating preferably over some distance? Now that we know that midges emit pheromones that invite legions of other midges to join in the feeding—can we disrupt this perhaps by wearing our own personal anti-invitation pheromone attached as a badge to blouse or jacket? Now that we have some information on swarms and copulation, can we develop ways of interfering with midge reproduction, particularly at a local level? After all, bee-keepers regularly use chemically-synthesized pheromones to lure their swarms. The prospects are there given the appropriate backing.

There is one area of research which perhaps has more long-term promise than any other of solving the midge problem. This is what is called 'biological control'. At its simplest this means the control (or even the elimination) of one undesirable organism by another, less undesirable, biological component. The biological component may be a predator introduced into an area because it specifically seeks out and kills the unwanted species. An example is the deliberate introduction of a cat on to a rat-infested island. We have moved on from the crude, often disastrous, biological controls of the past. Today, the biological component is often a disease, viral or bacterial, which is intentionally spread to bring about the demise of a pest species. An example is the deliberate spread of a particular virus which destroys the Pine Beauty Moth larva, a menace capable of devastating the forest plantations of Scotland. Other insect pests are being tackled, overseas, by the introduction of genetically engineered bacteria which are pathogenic to the pest in question and, seemingly, to no other. One such organism, *Bacillus thuringensis*, is a common soil bacterium which forms resting spores. These spores can be cultured commercially and sprayed on to the target pest. Inside the spores are crystals of toxic proteins which are then digested in the highly alkaline gut of the insect larvae. Because the digestive juices of each species of insect differs from other insects, the toxic crystals can be genetically modified and targeted to particular pests—at least in theory. Whether or not this pathogen would be effective against the Highland Midge has yet to be investigated. The bacterium has been used successfully under strictly controlled conditions and on a small scale against a localized population of Black Flies way down in Dorset.

It is also possible to transfer anti-midge genes cut from *B. thuringensis* into plants. There the genes express themselves in the plant and subsequently cause the death of insects which choose to feed upon the transformed leaves. This works well, for example in defending the tobacco plant against tobacco hornworm. To extend this powerful weapon against Scottish midges we need to discover what moorland plants are fed upon by the male and female midges. This relatively unsophisticated but essential research appears never to have reached the scientific literature (if it has ever been done).

Midges have their natural enemies, animal, fungal or bacterial, and it may be possible to use these to control the midge population. For example, midges harbour parasitic mites and probably have done since earliest times—as can be seen in mite-infested midges preserved in 70 million year-old amber. It is the immature mite which parasitizes the midge—the adult mite feeds on some unknown prey. It may be possible to promote mite infestation at midgey sites—though we are a long way from knowing when and how midges become infested. Indeed the importation of mites as a biological control would be thoroughly irresponsible without a major research effort to establish the specificity of the mites. The midges round the Tay and the Forth on the East Coast are sometimes infected by a minute parasitic worm. The same nematodes may, indeed, form part of the midge larvae diet—a case of eat or be eaten. Some of these nematode parasites appear to specialize in living inside the midge larvae without actually killing their host. Instead the poor old midge, when it becomes an adult, develops a mix of male and female sex organs and is effectively sterile. Field trials in California where nematodes have been targeted against midge eggs have shown a significant reduction in emerging adults. However, the American nematodes appear to have a catholic taste in insect larvae and related nematodes in Europe are known to be not choosy in their diet. Limited trials have been conducted in Scotland recently but again it would be highly irresponsible to liberate nematodes on a field scale until much more is known of their biology. Increasingly, biological control methods now have to be thoroughly vetted by government agencies charged with responsibility for safety in the environment.

To overcome the objection of introducing parasites as a control method, sterilization of immature males has proved a useful technique particularly against the Mediterranean Fruit Fly, aided by the fact that the male fruit fly pupae can be separated from the female simply by body colour. Once released these males mate with females which then fail to produce a new generation. However, as laboratory techniques for culturing the Highland Midge are in their infancy—let alone the problem of sexing the pupae—it may be some years before sterilization can be developed as a practical control method.

There is, however, one over-riding problem with biological control—the Highland Midge is hugely successful because there are millions of hectares available for breeding. Money invested on introducing biological weapons to localized sites will fall foul of the same problems exposed by the use of DDT forty years ago—adult midges from neighbouring upwind breeding grounds will drift passively into treated sites and continue biting.

The inevitable question that arises when science holds out prospects of a solution is who pays for the research needed to reach development? Funding for basic biological science, today, has to compete with many other demands from medicine, veterinary science, agriculture, ecology and environmental sciences. In Britain, science funding has to be justified also on the grounds of its worth in terms of quality of life and wealth creation. One solution has been that the customer should pay. The customers, those who stand to benefit from the research, would include the local Highland communities, the summer visitors whether as transient coach parties or longer-stay hotel guests, sportsmen including game-fishermen on the Spey and climbers on the Munroes. Others to benefit would include those who earn at least part of their living from the tourist trade whether as keepers of hotels, youth hostels, craft shops or from one of the many other businesses which thrive during the summer season. The Highland industries represented by forestry, fishing, fish farming, crofting, weaving and knitting would also benefit to some degree from midge control. In addition, other groups are concerned about biting midges but may live out of the region. These will include the medical and veterinary professions charged with health care in Scotland and the chemical and pharmaceutical industry involved in the manufacture, import and distribution of midge repellents, insecticides and related products. With such diffuse interests it is unlikely that any one sector would be willing to finance all of the research needed to alleviate the midge problem. There is then a need for a central authority to initiate and to promote research. Fifty years ago the initiative was taken by the then Secretary of State for Scotland with funding principally from the Department of Health for Scotland and the Scottish Tourist Board. Times and attitudes to community-wide problems have changed over the half-century but if the will is there to support the research, then new answers will be found to tackle the midge problem. It could benefit from a lead.

## *THE ECOLOGICAL SIGNIFICANCE OF MIDGES*

Whatever Man has tried to do to control the activities of midges, there remains the hard fact that midges in turn have been controlling Man's activities for many, many years. Whether or not this age-old struggle will continue in the midges' favour remains to be seen. But quite apart from their relationship with humans, the biting midges have an important and enduring relationship with many of the other creatures that go to make up the natural history of the Highlands.

At one level, midges, like most other insects, are part of a complex food web. The sheer number of midges, and particularly midge larvae, throughout the West Highlands are bound to make an impact on the lives of other organisms whether as prey or as a food-source. The role played by midge larvae in these food chains can only be guessed at, for this is another neglected area of scientific research. It is known, however, that the midge larvae are the food source for several types of soil-dwelling invertebrates, including nematode worms. Similarly, the omnivorous midge larvae prey heavily on other soil organisms as can be seen from their five- or six-fold increase in body length in nine or ten months. Because they are a component of one or more food chains, then any episode, natural or Man-made, which significantly alters the size of the midge larval population will inevitably have a knock-on effect on the other inhabitants of the soil. Droughts in early summer, for example, are known to reduce midge numbers drastically though what effect removal of midges has on other creatures is not known. In a similar way, we are ignorant of the effects of anti-midge larvicides, biological or chemical, on all those soil-borne creatures which directly or indirectly rely on midges as a food source.

Leaving aside the poorly understood world of the larvae, it is the impact made by the flying adult midge which has attracted Man's attention and more often his wrath. This impact lies at the heart of the ecology of the Highlands because it directly modifies Man's activities. In Eastern Russia, where biting flies are a serious pest, ecologists have shown that midges and other blood-sucking flies are probably essential to the maintenance of the fragile tundra by ensuring that the larger grazing animals keep on the move. Twenty years ago Boris Dubitskii described the role of blood-sucking insects in the ecology of the Siberian taiga—one of the least exploited biospheres on Earth. There, midges, along with other blood-sucking insects, maintain an environment where large grazing animals (and humans) are kept continuously on the move so reducing their potential to exploit and destroy the fragile ecology of the area. The enormous population of blood-sucking insects acts as a potent form of biological control—the control this time being exerted against large mammals (and incidentally, humans). In Scotland, the Highland ecology has already suffered much from several thousand years of human exploitation, but compared with the rest of Western Europe the destruction has been relatively limited in duration and degree and until recent centuries largely confined in intensity to the lower slopes of hills and to accessible straths and glens. One reason for this has been the activity of midges which has kept Man well in his place.

Midges breed in the high rainfall areas of the Highlands and depend on the high soil moisture for survival. In recent years there have been indications of changes in our climate, as a result of our industrial outpourings, which could have a significant impact on land-use and hence the wildlife of the Highlands. However, the impact of climatic change, particularly in the west, will depend greatly on future rainfall patterns, something climatologists have not been able to predict with any certainty. Current forecasts suggest that a global rise in mean annual temperature of 1 to 1.5°C. may occur over the next 50 years. To maintain the high soil moisture of the West Highlands at these temperature rises would require an increased rainfall of about 10 per cent. If there is no such increased rainfall then the extent of midge breeding sites might well diminish as the soil dries out. Such a prospect remains, for the moment, at best a long-term possibility.

Today, and for the foreseeable future, midges continue to breed over much of the Highlands in precisely the places which have long proved to be the most difficult to cultivate and maintain. Here both Man and his domestic animals have struggled to secure some sort of productive and ordered life for the past 4,000 years or so. Until recently this struggle was confined largely to the most accessible coastal strips. The broader expanses of moorland, woodland and high hills have long been considered to be, at best, half tamed. The principal reason for this is that the environment allows just a degree of exploitation before it responds by becoming even more harsh and economically less rewarding. The stripping of the natural forests from the lower hillsides in the past four millennia has left behind land which has fallen to the ever-encroaching blanket bog and ill-drained moorland. Into this impoverished landscape the Highland Midge has moved to become, today, a dominant member of the natural wildlife: far, far more significant than the golden eagle, pine marten or wild cat. Having achieved this dominance over the Highlands, the biting midge should be recognized as an important element in limiting human activities. Within the context of Western Europe, the Scottish Highlands are one of the most under-populated landscapes with a timelessness difficult to find anywhere in the late twentieth century. This is an area where Man has a foothold and no more. If, as seems likely, the biting midge is a significant factor in limiting our grossest activities then this diminutive guardian of the Highlands deserves our lasting respect.

# *BIBLIOGRAPHY*

There are few readily obtainable (and readable) books on biting midges, but for those who would like to delve into the scientific literature, a brief list is given below. It is generally possible to borrow the appropriate book or journal through the public libraries. The selection given can only be the tip of the iceberg, but at least it is the more readable tip.

1. Books with broad-based accounts of biting midges together with related insects and their impact throughout the world:

Kettle D S, (1995) *Medical and Veterinary Entomology*, Second edition, CAB International, Wallingford.

Service M W, (1986) *Bloodsucking Insects: vectors of disease*, Studies in Biology Series No. 167, Edward Arnold, London.

2. Scientific accounts in which the reader can delve further:

Blackwell A, Mordue A J, Young, M R and Mordue W (1992). Bivoltinism, survival rates and reproductive characteristics of the Scottish biting midge *Culicoides impunctatus* in Scotland. *Bulletin of Entomological Research*, **82**, 299-306.

Boorman, J (1986) British Culicoides (Diptera: Ceratopogonidae). Notes on distribution and biology. *Entomology Gazette* **37**, 253-66.

Campbell, J A and Pelham-Clinton, E C (1960) A taxonomic review of the British species of Culicoides. *Proceedings of the Royal Society of Edinburgh.* **B 67**, 181-302.

Edwards, F W, Oldroyd, H and Smart, J (1939) *British Blood-sucking Flies,* British Museum, London.

Hendry, G A F and Godwin, G (1988) Biting Midges in Scottish forestry. *Scottish Forestry* **42**, 113-19.

Jennings, D M and Mellor, P S (1988). The vector potential of British Culicoides species for blue tongue virus. *Veterinary Microbiology* **17**, 1-10.

Kettle, D S (1969) Ecology and Control of Blood-sucking Ceratopogonids. *Acta Tropica* **26**, 235-248.

Kettle, D S (1961) A study on the association between moorland vegetation and the breeding sites of Culicoides. *Bulletin of Entomological Research* **52**, 381–411.

Mallett, J (1989). The evolution of insecticide resistance: have the insects won? *Trends in Ecology and Evolution* **4** 336-340.

Roberts, A (1995). A rising cloud of midges in the Scottish Highlands? In *Ecological Relations in Historical Times* edited by R A Butlin and N Roberts, pp 88-98, Blackwell, Oxford.

# APPENDIX

## BITING MIDGE (CULICOIDES) SPECIES FROM SCOTLAND

| *Species* | *Frequency* | *Breeding sites* | *Host* | *Human biting* | *Comment* |
|---|---|---|---|---|---|
| *C. achrayi* | Uncommon | Loch margins | Horses | (+) | |
| *C. albicans* | Common | Wetter bog areas | — | | Biting habit lost? |
| *C. brunnicans* | Uncommon | Streamsides | — | (+) | Early season flier |
| *C. cameroni* | Rare | Waterfalls | — | | Argyll only? |
| *C. chiopterus* | Common | Cow dung | Cattle | (+) | Diminutive |
| *C. circumscriptus* | Locally common | Salt marshes | — | | |
| *C. clintoni* | Rare | Shaded bogs | — | | |
| *C. delta* | Common | Wet areas, moors | Cattle, horses | (+) | Early season flier |
| *C. dewulfi* | Common | Dung, stables | Horses | (+) | |
| *C. duddingstoni* | Local | Loch margins | — | | |
| *C. fascipennis* | Common | Wetland mud | Birds? | (+) | |
| *C. fagineus* | Uncommon | Damp soil, moors | — | | |
| *C. festivipennis* | Uncommon | Damp soil | Birds | | [synonym *C. odibilis*] |
| *C. grisescens* | Common | Marshes | Cattle | (+) | Overwinters as eggs |
| *C. halophilus* | Locally common | Coastal brackish pools | — | ++ | Painful biter in Gairloch |
| *C. heliophilus* | Frequent | Sphagnum bogs | Sheep, dogs | +++ | Daytime flier |
| *C. impunctatus* | Abundant | Wet acidic grassland | Cattle, deer, horse | +++ | The Highland Midge |
| *C. kibunensis* | Frequent | Marshes, loch margins | Horses, birds | | [synonym *C. cubitalis*] |
| *C. lupicaris* | Uncommon | Exposed mud | Cattle, horses | (+) | |
| *C. manchuriensis* | Uncommon | Salt marshes | — | | Biting habit lost? |
| *C. maritimus* | Locally common | Salt marshes | — | | |
| *C. nubeculosus* | Common | Farms, fanks | Horse, cattle, sheep | +++ | Vector of *Onchocera cervicalis* in horses |

| | | | | | |
|---|---|---|---|---|---|
| *C. obsoletus* | Abundant | Garden compost | Cattle, sheep | +++ | Mainly Lowlands |
| *C. pallidicornis* | Frequent | Marshes, ditches | Horses, birds | (+) | Especially Lowlands |
| *C. pictipennis* | Common | Woodland pools | Birds | (+) | Early season flier |
| *C. poperinghenis* | Rare | Tay salt marshes | — | | |
| *C. pulicaris* | Locally common | Muddy marshes | Horses, cattle, sheep | +++ | Sweet itch of horse |
| *C. punctatus* | Frequent | Muddy marshes | Horses, cattle, sheep | + | Sweet itch of horse |
| *C. reconditus* | Uncommon | Not known | — | | |
| *C. riethi* | Locally common | Forth, Tay salt marshes | Horses | | Daytime flier |
| *C. riouxi* | Rare | Not known | — | | Strathclyde |
| *C. salinarius* | Uncommon | Salt flats | — | | Common on Beauly Firth |
| *C. scoticus* | Rare | On fungi?, wetlands | — | | |
| *C. segnis* | Uncommon | Not known | — | | |
| *C. stigma* | Common | Exposed mud | Horses | + | |
| *C. vexans* | Locally common | Damp soil | — | ++ | Suburbs, enters houses |

*Key:* Biting habit: +++ = Persistent biter of Man, + = less frequent biter of Man, (+) occasionally bites Man.
— = most common biting habits in Scotland not known